Cultivating and Communicating Natural and Technical Knowledge around 1800

Cultivating and Communicating Natural and Technical Knowledge around 1800

Devotees of Science

Edited by
Nikolaj Bijleveld,
Arjen Dijkstra, and
Samuel Gessner

Routledge
Taylor & Francis Group
LONDON AND NEW YORK

First published in 2025 by Amsterdam University Press Ltd.

Published 2025 by Routledge
4 Park Square, Milton Park, Abingdon, Oxon OX14 4RN
605 Third Avenue, New York, NY 10158

Routledge is an imprint of the Taylor & Francis Group, an informa business

ISBN: 9789463728577 (hbk)
ISBN: 9781041177791 (pbk)
ISBN: 9781003693482 (ebk)

NUR 685

Cover illustration: Demonstration of the electrostatic generator in the Hall of Physics in the building of the Felix Meritis in Amsterdam, inaugurated in 1789. Engraving by Reinier Vinkeles, 1801 (*Collection Rijksmuseum Amsterdam, RP-P-OB-70.321*)

Cover design: Coördesign, Leiden

DOI 10.5117/9789463728577

Table of Contents

Mediation of Knowledge: Artefacts and Context

1. Devotees of Science: An Introduction

Nikolaj Bijleveld, Arjen Dijkstra, and Samuel Gessner

Abstract: Following historiography on the development of natural knowledge and its relation to the Enlightenment and Romanticism, the introduction situates the present volume in the perspective of an ongoing research endeavour aimed at understanding the field of knowledge advancement in the decades around 1800. The protagonists of this volume are presented, and a proper term is introduced – devotees of science – to designate these persons of diverse professions and occupations who could and wanted to devote their spare time and personal resources to advancing scientific knowledge. The calling of men and women to become a devotee of science, and its critique – usually voiced by elite scholars – was in tune with the Enlightenment ethos and not infrequently with rising national awareness. The devotees' role in the second half of the eighteenth century must be understood against the background of traditional scientific institutions reforming and new institutions emerging. Finally, the structure as well as the contributions to the volume are introduced.

Keywords: knowledge advancement, cultivation of knowledge, Europe, Enlightenment, Romanticism

Just before the close of the eighteenth century, two prominent entrepreneurs from the northern region of the Netherlands, Yme Freerks Tigchelaar (1753–1812) and Eise Eisinga (1744–1828), formulated a comprehensive strategy to rescue their local university in Franeker from decline.[1] They were commissioned by the regional government, which was looking for arguments to keep allocating funds to the university while at the same time opening it up

1 Tigchelaar and Eisinga, *Plan*. See more on this *Plan* and on the writers in the contributions by Arjen Dijkstra and Fokko Jan Dijksterhuis to this volume.

Bijleveld, Nikolaj, Arjen Dijkstra, and Samuel Gessner, eds. *Cultivating and Communicating Natural and Technical Knowledge around 1800: Devotees of Science*.
Taylor & Francis Group, 2025
DOI 10.5117/9789463728577_CH01

to a broader audience. In their report, Tigchelaar and Eisinga outlined the novel guiding principle of the university that advancement of knowledge as well as of the natural sciences depended on practical activities, ultimately benefiting society as a whole.

Tigchelaar and Eisinga were not professors, nor had they ever studied at a university. They were not themselves members of any scientific society or institution, nor were they noblemen with traditional authority. They were locally renowned for their accomplishments: Tigchelaar for running a well-known porcelain factory and Eisinga for his successful wool trade, and even more for his design and construction of the Franeker Planetarium – the largest and most complex orrery of its time. While neither had held any institutional scientific role, both were, thanks to their social position and intellectual achievements, deemed the appropriate men to provide advice on the way forward.

A significant number of actors played an important role in the production and communication of knowledge in Europe around 1800 but nevertheless slipped through the conceptual grid of historiography. This happened to those whose personae could not be identified *a posteriori* with what the nineteenth century would start to call 'scientists'. Who were these persons who cultivated knowledge of the natural world and mastered scientific apparatus outside traditional institutions? The goal of this volume is to pursue that question by looking at the practice of those involved regardless of their status and historiographic representations. Its aim is to find a proper place in history for these people, who we will refer to as 'devotees of science'. It is also a step towards reshaping the narrative of eighteenth-century science by highlighting the role of actors who were not part of traditional institutions. Finally, it is hoped the nature of this volume, a series of case studies, will open pathways for additional research that follows the same direction.

Our Question: The Bias of Institutional Science

The devotees of science can be found all across Europe around 1800. Historians of science venturing outside the eighteenth-century locales where scientific activity is retrospectively expected to be found will often encounter persons of diverse professions and occupations who could and wanted to devote their spare time and personal resources to advancing scientific knowledge through practical means, crossing geographical, linguistic, and societal barriers. They contributed considerably to the spread of new ideas and found ways to contribute to new developments.

These actors have not been entirely neglected by historiography so far. Their treatment, however, has remained inadequate for two reasons. Firstly, because many have never been affiliated with any institution, nor were they among the authors of memoirs of academies. Therefore, they were considered marginal figures of little importance. Secondly, individual cases may have received much attention but were subject to questionable interpretation, for instance by being overly romanticised. These persons were portrayed as exceptional in both senses: as amazing and atypical.

One of the premises of this book is that historically these actors were neither marginal nor atypical. Their number was large, and they had a broad variety and historical role worthy of renewed and greater consideration. They were engaged in innovating, disseminating, accommodating, and presenting all kinds of knowledge, be it scholarly and of academic origin or more practical or technical. They did so, often by focusing on the practical, spurred by a conviction to personally contribute to the enlightenment process that involved social and economic improvements. They often found their way into newly founded institutions, young academies and societies that increasingly competed with existing locales of knowledge (courts, universities, religious colleges). It is also true and even characteristic of them that an academic training and a good control of Latin were deemed less important. Furthermore, they made a living that did not totally align with their engagement in the sciences. Conspicuously, however, their cultivation of knowledge of nature and technical expertise won them cultural capital that served as a lever to aid their rise in society.

Their multiple roles, therefore, need to be more fully studied in recognition of the many instances of influence they had on the transformation of the scientific landscape in Europe by the close of the eighteenth century. Traditionally, the institutions brokering formal and academic knowledge constituted spaces reserved for the elite. The social privilege of science, however, came under pressure in the eighteenth century as the cultivation of knowledge was no longer limited to the upper classes. Enlightenment ideology postulated that knowledge should be spread and put to use. Consequently, works were translated and published in the vernacular and made accessible through compendia and popularised editions. New institutions emerged and traditional ones underwent reforms, while a new social class emerged and found an opportunity to distinguish itself.[2]

Contextualising our protagonists is essential, not only for making sense of their individual actions, but also to gain a fuller understanding of the

2 See Klaas van Berkel's contribution to this volume.

processes driving the history of science and academia. Before we discuss how 'devotees of science' have been understood, we need to take a closer look at the context within which this was done: that of science and academia at the close of the eighteenth century.

State of the Problem: Science and Academia around 1800

Research on scientific cultures up to and around 1800 has seen a lot of illuminating contributions over the last few decades. The general trends in historiography that increasingly inscribed scientific developments in the economic, political, cultural, and ideological contexts have done much to fertilise these endeavours. To name but a few of the decisive impulses of this research field we may briefly recall the following four moments.

Daniel Roche's massive two-volume work on the academies and 'academicians' in the French provinces made exceptionally clear the value of looking beyond the places that counted as the unquestioned centres of scientific progress and enlightenment.[3]

A few years later, in 1982, Geoffrey Cantor described the challenges of studying science in the eighteenth century with these words: The period 'seems devoid of both giants and key ideas'.[4] It was this lack of a 'cogent synthesis' that made it so hard to study. He coined this 'the eighteenth-century problem'.[5] Cantor's message resonated so strongly that it seems that in the years that followed almost everybody moved beyond this observation. What he coined as a problem paved the way for numerous subsequent studies that began to unveil the depth of scientific endeavour during the Enlightenment. This proved to be the beginning of an understanding of the process that was taking place: It was not the 'giants' or 'key ideas' that helped with this, but the 'institutions, geopolitics and ways of knowing' that aided in understanding and studying science in the period.[6]

Less than a decade after Cantor, in 1990, Andrew Cunningham and Nicholas Jardine published their work *Romanticism and the Sciences*, focusing on the study of science in the period immediately following the eighteenth century. They too moved far beyond the search for iconic figures, and they

3 Roche, *Le siècle*.

4 See the recent special issue by Bycroft and Wilson of the *Journal of Early Modern Studies* 12, no. 1 (2023), on the reception of and follow up on Cantor.

5 Cantor, 'Essay Review', p. 59.

6 Bycroft, 'Introduction', p. 11.

argued that the practice of science in the early nineteenth century displayed significant continuity with the preceding era. They saw numerous instances in Romanticism where scientific inquiry endured, which was dependent on 'the preservation of practices and agendas of inquiry [rather] than on the maintenance of doctrines'.[7] In that same volume, David Knight argued that 'Romanticism was [...] not a polar opposite of natural science as some have supposed', challenging the conventional view of a stark division between the two.[8]

In the years after the 1990s this way of practising history of science gained traction, but also raised new questions and brought new areas into focus. An important contribution came with the publication of *The Mindful Hand* by Lissa Roberts, Simon Schaffer, and Peter Dear.[9] This work explored the entire early modern period and moved well into the nineteenth century. The authors sought to elucidate how in historiography 'head and hand were put so strikingly at odds', while they did not find such a dichotomy between the two – they found a close relationship between the useful (hand) and new knowledge systems (head).[10] They also addressed how the public at large 'made sense of changes in the public status of skill and labour' around 1800.[11] This revealed how society started to comprehend the changes in the 'ways of knowing'. This was a process in which individuals outside the academic circle, our devotees of science, actively disseminated knowledge originating from within it, while at the same time influencing what happened inside that circle. This volume looks again at individuals, though not the giants and not even well-known figures from the mainstream narratives of the history of science.

In sum, as historiography has amply shown, during the period around 1800, universities, academies, and learned societies experienced substantial transformations.[12] The period was characterised by the establishment of new academies and learned societies, efforts to improve society through knowledge, reforms in economic methods, and an enthusiasm for sharing and spreading new inventions. Many of these institutions, reforms, and inventions have been studied individually or as part of broader syntheses. From there it appears as a natural step to identify the whole range of those

7 Cunningham and Jardine, *Romanticism*, p. 7.

8 Knight, 'Romanticism and the Sciences', p. 22.

9 Roberts et al., *The Mindful Hand*; see also Klein and Spary, 'Introduction', p. 3, who continued the work by Roberts et al.

10 Schaffer, 'Introduction', p. 323.

11 Schaffer, 'Introduction', p. 310.

12 See Klaas van Berkel's contribution to this volume.

who contributed to science, and to pursue the path initiated by Cantor, again revealing the depth and breadth of the past. If we prolong the move proposed by Cunningham and Jardine, we may see the entire variety of persons whose activities ensured the continuity of practices in Romantic science.

In this sense, a lacuna in the rich available historiography appears to require filling, for which a proper concept is lacking. If we aim to construct a narrative that spotlights individuals who do not form a cohesive group but are integral to the larger developments of the era, then we need to find a way to pinpoint them. How can we highlight the role of such actors who were not part of any traditional institution of knowledge, yet effectively made use of existing structures? The case studies in this volume show that such individuals played a crucial role in advancing the practices and research agendas of their time. Historically, their activities constituted a noteworthy scientific practice during this period although not entirely novel nor to completely disappear later. Moreover, we claim that their significance has never been more momentous than around 1800.

Approach: Devotees, Amateurs, and Scientists

To approach the study of these historical actors, we choose a simple and straightforward method. To enhance their visibility we apply the designation 'devotee of science' as a single concept, despite the variety of their individual profiles. Meanwhile, the studies in this volume will collectively offer a more nuanced understanding of these individuals than simply contrasting them with the professionalisation of science.

Our choice of the term 'devotees of science' requires more explanation. By using 'devotee' we deliberately keep a certain distance from the growing body of literature that explores the role of amateurs in the nineteenth century.[13] That discussion takes place within the context of the overall professionalisation of science. The term 'amateur' was employed in the eighteenth and nineteenth centuries, and like 'liefhebber' and 'dilettante', it referred to people deeply engaged in a specific topic. Contemporary use of these terms often implies low quality or lacking seriousness in relation

13 To quote just a few: Guillemain and Richard, 'Towards a Contemporary Historiography', pp. 201–37, together with the various case studies presented in that special issue of *Gesnerus*; Knight, 'Science and Professionalism', pp. i, 53–67; Alberti, 'Amateur and Professionals', pp. 115–47; Chapman, *Amateur Astronomer*.

to professionals, and is therefore prone to misunderstanding and often comes with commentary on the problematic nature of the terminology.[14] However, opinions on this complex process vary.[15] Nearly half a century ago, Rousseau and Porter even spoke of an 'amateurization' of science in the eighteenth century.[16] They argued that 'natural history is the creation par excellence of amateurs, and there is nothing incompatible about the combination of amateur and high-quality science'. Similarly, in his book *Putting Science in Its Place*, Livingstone highlights the significant reliance on amateur men and women in field sciences, from archaeological digs to botanical surveys.[17]

Today, as the distinction between 'high-quality' science and other forms of scientific culture is no longer our main concern, we seek new terms. Allan Chapman introduced the term 'Grand amateurs' to refer to individuals such as Francis Baily or John Herschel, who had their own observatories and published their research while maintaining other sources of income.[18] But it remains questionable whether there should be a distinction between 'Grand amateurs' and 'amateurs' in general. Other terms were used, e.g. 'cultivators of science', to describe individuals who had achieved some level of merit in a specific field of knowledge.[19] Hsiang-Fu Huang summarises the various designations used in historiography, ranging from 'individual enthusiasts' to 'gentlemanly specialists'.[20] Jacques Roger, in an essay on eighteenth-century natural history, provided many examples of individuals who fall into such categories, including pastors, soldiers, and public servants. Some remained 'amateurs' throughout their lives; others became 'professional scientists' by entering an academy or receiving a royal appointment. We, therefore, think that a less ambiguous term is needed to encompass those we discuss in this volume: persons of diverse professions and occupations who devoted their spare time and personal resources to advancing scientific knowledge.[21]

14 Cannon, *Science in Culture*, p. 141. See for instance: Snelders, 'Professors', pp. 308–23, this survey of the progress of Dutch chemistry uses the term amateur only in the title, but recognises the importance of 'gentlemen', free-lance lecturers, and 'devotees' active outside the universities. Sliggers, 'Natuurkundige amateurs', pp. 67–102.

15 Barton, '"Men of Science"', pp. 73–119.

16 Rousseau and Porter, *Ferment of Knowledge*, p. 4.

17 Livingstone, *Putting Science in Its Place*, pp. 42–43.

18 Chapman, *Amateur Astronomer*, p. 50.

19 Devoy, 'Lunar Crater Models', pp. 300–32, especially p. 303. Devoy mentions the foundation of the British Association for the Advancement of Science in 1831 'to promote the intercourse of the *cultivators of science* with one another' (our italics).

20 Huang, *Commercial and Sublime*, pp. 44, 78.

21 Roger, 'The Living World', pp. 260–63.

According to current usage, 'devotee' refers to someone who is strongly committed to a specific topic. It is semantically similar to terms such as 'liefhebber', 'amateur', and 'dilettante'. We chose 'devotee' because it avoids the belittling connotations that these other terms may convey today. During the period under consideration, around 1800, the term was not commonly used, and we use it retrospectively. Since this volume explores the position of these devotees within the changing landscape of academia and learned societies, as well as their relationship to the development of science and scholarly culture, we decided to avoid adjectives such as 'non-scholarly' and 'non-academic', or the label 'practitioner'. These terms imply an opposition, whereas usage of the term 'devotee of science' allows for a discussion of the relationship between these individuals and formal educational and knowledge institutions.

Additionally, when the term 'scientist' is used in the following chapters, it is done so cautiously. This term came into use in the nineteenth century to describe those who primarily studied natural sciences at academic institutions. Since the figures in this volume, at most, contributed to the establishment of these natural sciences through their experiments and empirical methods, they were primarily engaged in natural philosophy. However, the terms 'science' and 'scientist' are still appropriate translations for the German words 'Wissenschaft' (and the Latin term 'scientia' and Dutch term 'wetenschap') and 'Wissenschaftler'.

As a loose cut-off for the contributions to this volume, we have chosen the professionalisation of the sciences in universities and the separation of the humanities from the natural sciences in the nineteenth century. Some contributions discuss the early eighteenth century, when the ideal of attaining universal knowledge began to be replaced by the practical application of knowledge or, more cautiously put, when natural knowledge played a significant role in the emerging sphere of 'public opinion'. Institutionalised sociability and the rise of periodicals, for example, paved the way for the devotees of science.[22]

Result: Devotees in Various Constellations – Structures and Themes

This volume presents the world of the devotees of science through the lens of three overarching themes, all of which are discussed in the subsequent

22 Burke, *The Polymath*, pp. 83–125.

chapters. As the initial inspiration derived from a workshop about Eise Eisinga, it is appropriate that we begin with a section that places Franeker centre stage, as indicated by the title of the first section, 'Cultivators of Knowledge: From Europe to Franeker'. The opening chapter, by Klaas van Berkel, links Eisinga's hometown to the European context, providing insight into the framework in which devotees of science around 1800 thrived. Van Berkel asserts that Eisinga was 'discovered and utilised by a university professor' and underscores the 'extensive contacts' of Professor Van Swinden (1746–1823) within the academic realm, which were pivotal for the Franeker Planetarium's instant success. The chapter highlights how devotees of science, in general, were closely connected to established scholars and institutions, even if they themselves were not part of academic and scientific circles.

With the stage set by Van Berkel's chapter, Arjen Dijkstra's contribution provides a detailed analysis of the Franeker Planetarium and its creator. While Eisinga has often been portrayed as a remarkable figure with exceptional mathematical skills, Dijkstra argues that it is more accurate to see him as a successful, politically active entrepreneur who was also passionate about astronomy and mathematics. This perspective highlights the societal importance of Eisinga's planetarium, and the prestige associated with working in the fields of astronomy and mathematics. His efforts received widespread attention, and the news of his planetarium rapidly spread worldwide, making it a window into scientific practices elsewhere.

The second theme we address in this volume delves into how these devotees of science harnessed their unique skills, knowledge, and access to scientific practices. The four chapters under the heading 'Brokering Knowledge' explore individuals from across Europe who leveraged their skills and gained access to new knowledge and networks. Often, knowledge-sharing was aimed at the margins or lower social classes, but it is crucial to note that these intermediaries also facilitated a flow of knowledge from the periphery to the centres of intellectual prestige. Status played a significant role for these intermediaries in various ways, as the success of their contribution to knowledge creation was determined not only by their intellectual competences. Their social skills as well as social position were at least as significant in achieving recognition for their contribution.

Hochadel's account of the Austrian itinerant lecturer Martin Berschitz (dates unknown) illustrates how knowledge from natural philosophy was shared across space and social classes. Berschitz reached diverse audiences, piqued interest in science, and adapted his methods. He strived for access to the knowledge organisations that propagated the Enlightenment ideal but faced challenges and criticism from the scholarly community that concerned not only his knowledge and activities. The way in which his alleged lack of civilisation was invoked to keep many doors closed reveals the role social criteria played in the history of science during this phase and explains why Berschitz's contribution to it remained modest compared to those of contemporaries who were included in the knowledge infrastructure of the time.

Gluch introduces the 'paysans' Christian Gärtner (1705–1782) and Johann Georg Palitzsch (1723–1788) as examples of commoners in the late eighteenth century who defied societal expectations to pursue knowledge. They built local networks that supported their scientific ambitions, particularly in the study of comets, which provided opportunities for recognition beyond their social ranks. Gärtner faced criticism for his perceived lack of knowledge, whereas Palitzsch embodied the ideal of an educated commoner, combining theoretical scientific knowledge with humility and diligence in his agricultural work. These cases reveal how curiosity, resourcefulness, and social networks shaped visibility and reputations in a stratified society.

The French officer Alexandre Moreau de Jonnès (1778–1870), who found his niche during the transition from the old educational system to the professionalisation of the sciences, was in a similar situation. As an illegitimate child with contested claims to nobility, he pursued a professional career while dedicating himself to the study of statistics, geology, zoology, and more. Unlike Berschitz, Moreau de Jonnès sourced and produced most of his knowledge outside the upper echelons of society. Yet, as the authors Hugues Chabot, Pierre Crépel, Thierry Dumont, Christian Gaillard, Maryannick Lavigne-Louis, François Renaud, Denis Reynaud, Dominique Saint-Pierre, and Thomas C. Anderson argue, he successfully connected with the established intellectual infrastructure in France.

Nikolaj Bijleveld shows in his contribution the active engagement of Danish and Dutch pastors in advancing and disseminating knowledge beyond their religious roles. These pastors pursued scientific interests, conducted research, and published works that contributed to individual and societal prosperity. As true brokers, they facilitated the exchange of new knowledge between the (rural) peripheries and the centre. The comparison between these pastors in Denmark and the Netherlands highlights the factors that led to their success. Their aspiration, ability to apply knowledge, education, and network access enabled them to contribute to an international ideal of educating the people and enhancing national and social prosperity.

The third theme of this volume partly follows the approach of Siskin and Warner, who view the Enlightenment through the lens of media studies.[23] They suggest considering the Enlightenment as a historical event centred around mediation. This perspective offers several advantages. It highlights key elements of the Enlightenment, including new communication infrastructures, and the proliferation and eventual saturation of mediation, all closely linked to the rise and fall of the Enlightenment movement. Where do devotees of science fit into this view of the Enlightenment as a mediation event? The chapters in the final section, 'Mediation of Knowledge', aim to address this question.

Certainly, devotees played a crucial role in creating new mediation infrastructures, alongside academies, public lectures, periodicals, and popular print publications. While not all of them acted as 'grand amateurs' or 'gentlemanly specialists' who fostered international networks within the Republic of Letters, most were essential for the widespread circulation of knowledge in society. Clearly, during the Enlightenment, the systematic organisation of knowledge, such as large encyclopaedia projects, was a deliberate effort at knowledge mediation. Along the same lines, the growing industry and diverse technology for visualising knowledge became a prominent mode of mediation.

As Wouter de Vries's contribution demonstrates, mediating knowledge about the supralunar space through illustrations led to evolving perspectives over time. Using visual elements in knowledge mediation set in motion changes in the knowledge itself. This chapter also shows how these illustrations not only shaped popular understanding of astronomical knowledge but also guided the research of the heavens. Mechanical models, such as orreries, were a particular focus for many cultivators of science and showcased the enthusiasm of the time for this new medium. The chapter provides a novel insight into the context in which the devotees operated.

Luís Tirapicos and Samuel Gessner's chapter shows how the Oratorian priest Teodoro de Almeida (1722–1804) consistently sought innovative forms of three-dimensional representation for his teaching, consciously adapting his models to engage specific audiences. Working in Portugal, the Basque Country, and southern France, Teodoro exemplifies the international nature of this endeavour and how the new forms of knowledge transcended borders of countries, languages, and class.

The progress of knowledge depended on critical examination of existing knowledge. Therefore, as Fokko Jan Dijksterhuis's chapter demonstrates,

23 Siskin and Warner, *This Is Enlightenment*, pp. 10–21.

accessing the knowledge of various crafts and arts became a fundamental pursuit. Many science enthusiasts were writers and editors who meticulously collected knowledge and conveyed it through new literary genres. This underscores the feedback loop between simulating the universe through a planetarium, comprehending emerging forms of knowledge, and the process of asking new questions.

Preliminary Conclusions: An International Phenomenon

The goal of bringing these various cases together in this volume is to gain new insights into the context in which these science enthusiasts operated, what characterises them and where their contribution to the further development of knowledge and science lies. The first conclusion we can draw from this exercise is that there is a lot of work that still needs to be done. There is a multitude of devotees that can help to advance our understanding further.

Many of our devotees have been portrayed as self-taught, reflecting a prevailing sentiment that in the late eighteenth century universities were obsolete institutions. However, this view is biased and incomplete, as for example, Arjen Dijkstra argues in this volume. A closer examination of the careers of devotees reveals that they often received training from private tutors and certainly did not operate in isolation. Education is just one aspect that helps us understand these individuals and the broader context of their era. It is crucial to consider their networks, connections to educational and professional institutions, as well as the motives behind their pursuits.

The dissemination and expansion of knowledge about the natural world during this period were not driven solely by institutions such as academies, learned societies, and schools. Private initiatives played a significant role as well. This volume explores various examples of devoted individuals from different professional backgrounds who were interested in various scientific disciplines. These devotees often emerged from the new middle classes and combined their passion for knowledge with a sense of patriotism, nation-building, moral education, and socialisation. Moreover, these private initiatives contributed to the professionalisation of science through their involvement in the establishment of universities and scientific disciplines. Many higher education institutions were establishing their identity during the late eighteenth and early nineteenth centuries, and their endeavours were seen as platforms for the public dissemination of knowledge and national self-awareness.

The actions and works of the devotees are documented in various artefacts, designs, museums, popular books, and archival records that have been preserved. These sources not only reveal the practical aspects of their work but also shed light on their motivations and the support they received.

The case studies in this volume examine how these devotees and their activities connect to contemporary developments at different levels – epistemological, political, social, and cultural. Science was practised by individuals and groups from various social backgrounds and professions across Europe. The chapters present a range of protagonists, including pastors, merchants, agricultural and factory managers, soldiers, and freelance consultants who also acted as entertainers. Together, these case studies paint a variegated picture of how devotees fit into the broader knowledge-production landscape in the eighteenth century. Some of the individuals featured in this volume, despite their clear ties to the church, did not explicitly express their commitment to faith in their scientific pursuits. But faith often played a vital role in scientific development, and many contemporaries saw a natural connection between religion and science.

By examining these individual cases side by side, we gain a better understanding of the environment in which these devotees thrived in their pursuits. Although a devotee of science was free and independent, his success required not only knowledge but also connections to established institutions such as academia and societies. These institutions provided access to knowledge and served as networks and platforms for sharing ideas and work. But the devotees also shaped and contributed to the new knowledge infrastructures that supported their ideas and facilitated their activities. Institutions disciplined the devotees by providing a normative enlightened framework for their activities and behaviour. Recognition of achievements and virtues as well as validation by authorities in the field and representatives from higher classes, as illustrated in the case studies presented by Oliver Hochadel and Sibylle Gluch, were essential for the success and position of the devotees. The successful pursuit of knowledge in line with Enlightenment thought contributed to one's social position. In that sense, the history of the devotee of science is as much the history of the rise of the bourgeoisie.

Additionally, intrinsic motivation and external factors such as education, networks, and financial independence influenced the activities of these devotees. Contextual factors were also significant. For example, Teodoro de Almeida's case highlights how individual talent and quality were shaped by the audience they aimed to reach. Similarly, the activities of clergymen in northern Europe were often dictated by the resources and demands of their parishes, with an emphasis on practical knowledge such as agronomy. The Enlightenment call for useful knowledge nurtured the dedication of these individuals. The context not only selected and inspired talented citizens as devotees but also guided and confirmed their interests. While these devotees were unique in their individual contexts, they were part of a broader European phenomenon.

This volume represents the initial step towards a more systematic exploration of individuals who devoted their time and resources to scientific endeavours. We employ the framework of the 'devotees of science' and provide a historical context for these individuals. The results are promising, and it is undeniable that numerous other individuals of both genders, not only in Europe but also beyond the geographical boundaries of this continent, pursued similar goals with similar effects. As many more individuals outside the scope of this volume are investigated, our understanding of the remarkable continuity and profound changes that occurred in the natural sciences around the year 1800 will only deepen.

The concept of the devotee of science as someone who, in addition to their own professional work, distinguishes themself in the field of knowledge advancement and dissemination fills a gap in the history of science left by the relative weakness of traditional knowledge institutions during this period and the tendency to focus on 'giants'. The devotees connected 'head' and 'hand'. They produced and became embedded in the new knowledge infrastructure of this era before the professionalisation of science would again marginalise them. Contemporary reactions to individuals such as Berschitz and Palitzsch, who were not included in this group, illustrate how the knowledge infrastructure and access to it was arranged. These cases say less about the actual impact of their knowledge practices in society and more about how the institutionalisation of the sciences took place.

The descriptions of the individuals in the various chapters of this volume thus provide the elements to circumscribe what devotees of science were and what they were not. They also demonstrate that the times were changing: the ideal of meritocracy was progressively imposed and both the ancient and the new regimes began to reckon with the expertise of the devotees. We think that these chapters will help to understand that the devotees of science did not merely exploit a niche in the scientific landscape of their time, but that their cumulative efforts left an indelible mark on its deep transformation.

Bibliography

Alberti, S.J.M.M., 'Amateur and Professionals in One County: Biology and Natural History in Late Victorian Yorkshire', *Journal of the History of Biology* 34 (2001), pp. 115–47.

Barton, R., '"Men of Science": Language, Identity and Professionalization in the Mid-Victorian Scientific Community', *History of Science* 41 (2003), pp. 73–119.

Burke, P., *The Polymath: A Cultural History from Leonardo da Vinci to Susan Sontag* (New Haven 2020).

Bycroft, M., 'Introduction: Science Beyond the Enlightenment', *Journal of Early Modern Studies* 12, no. 1 (2023), pp. 9–31.

Bycroft, M., and A. Wilson, ed., *Journal of Early Modern Studies* 12, no. 1 (2023) [special issue: *The Eighteenth-Century Problem, Forty Years On*].

Cannon, S.F., *Science in Culture: The Early Victorian Period* (Folkestone 1978).

Cantor, G., 'Essay Review. The Eighteenth Century Problem. The Ferment of Knowledge: Studies in the Historiography of Eighteenth Century Science', *History of Science* 20 (1982), pp. 44–63.

Chapman, A., *The Victorian Amateur Astronomer: Independent Astronomical Research in Britain 1820–1920* (Chichester 1998).

Cunningham, A., and N. Jardine, *Romanticism and the Sciences* (Cambridge 1990).

Devoy, L.E., 'Lunar Crater Models', *Nuncius* 35, no. 2 (2020), pp. 300–32.

Guillemain, H., and N. Richard, 'Towards a Contemporary Historiography of Amateurs in Science (18th–20th Century)', *Gesnerus* 73, no. 2 (2016), pp. 201–37.

Huang, H.-F., *Commercial and Sublime: Popular Astronomy Lectures in Nineteenth Century Britain* [Doctoral thesis] (University College London 2015).

Klein, U., and E.C. Spary, 'Introduction: Why Materials?', in: U. Klein and E.C. Spary, ed., *Materials and Expertise in Early Modern Europe: Between Market and Laboratory* (Chicago 2010), pp. 1–23.

Knight, D., 'Science and Professionalism in England, 1770–1830', *Proceedings of the XIV International Congress of the History of Science* (Tokyo 1971), pp. 53–67.

Knight, D., 'Romanticism and the Sciences', in: Cunningham and Jardine, *Romanticism*, pp. 13–24.

Livingstone, D.N., *Putting Science in Its Place: Geographies of Scientific Knowledge* (Chicago 2003).

Roberts, L., S. Schaffer, and P. Dear, *The Mindful Hand: Inquiry and Invention from the Late Renaissance to Early Industrialisation* (Amsterdam 2007).

Roche, D., *Le siècle des lumières en province académies et académiciens provinciaux: 1680–1789* (Paris 1978).

Roger, J., 'The Living World', in: G.S. Rousseau and R. Porter, ed., *The Ferment of Knowledge* (Cambridge 1980), pp. 255–83.

Rousseau, G.S., and R. Porter, ed., *The Ferment of Knowledge* (Cambridge 1980).
Schaffer, S., 'Introduction', in: Roberts, *The Mindful Hand*, pp. 309–24.
Siskin, C., and W. Warner, ed., *This Is Enlightenment: An Invitation in the Form of an Argument* (Chicago 2010).
Sliggers, B.C., 'Honderd jaar natuurkundige amateurs te Haarlem', in: A. Wiechmann and L.C. Palm, ed., *Een elektriserend geleerde, Martinus van Marum, 1750–1837* (Haarlem 1987), pp. 67–102.
Snelders, H.A.M., 'Professors, Amateurs, and Learned Societies: The Organization of the Natural Sciences', in: M.C. Jacob and W.W. Mijnhardt, ed., *The Dutch Republic in the Eighteenth Century: Decline, Enlightenment, and Revolution* (Ithaca 1992), pp. 308–23.
Tigchelaar, Y.F., and E. Eisinga, *Plan tot herstel der Academie te Franeker, door het provinciaal bestuur van Friesland* (Franeker 1798).

About the Authors

Nikolaj Bijleveld is managing director of the University of Groningen Business School.

Arjen Dijkstra is director of Tresoar; literature museum, archive and library of Friesland.

Samuel Gessner is an associate researcher at the Department of History and Philosophy of Sciences at the University of Lisbon.

Cultivators of Knowledge

From Europe to Franeker

2. The Changing Landscape of Science in Eighteenth-Century Europe

Klaas van Berkel

Abstract: In the late eighteenth and early nineteenth centuries, the scientific landscape in Europe underwent a major transformation. Universities were in decline, while scientific societies were expanding. Some universities had to close their doors, while others implemented or tried to implement major reforms, in particular by promoting useful knowledge. A growing number of devotees operated outside these institutions, but it is often forgotten that in order to gain recognition, they were still dependent on the goodwill of university professors and members of the academies to act as intermediaries. A striking example of this interdependence is the way in which Eise Eisinga's extraordinary planetarium became famous through a booklet written by the Franeker professor of philosophy and physics Jean Henri van Swinden (1780).

Keywords: decline and reform of universities, expansion of academies, Eise Eisinga, Jean Henri van Swinden

In February 1780, Jean Henri van Swinden (1746–1823), professor of physics at the University of Franeker in the Netherlands, paid a visit to Eise Eisinga (1744–1828), the Franeker wool comber who had just built an intricate planetarium in his house not far from the university. Van Swinden had been professor at the University of Franeker since 1767 and had become a well-known physicist (Illustration 1). In 1777 the Academy of Sciences at Paris had awarded him and Charles Coulomb (1736–1806) a gold medal for a prize essay on the variability of the magnetic needle, and in 1778 he had received another gold medal, awarded by the Bavarian Academy of Science of Munich, for an essay on the analogy of electricity and magnetism. Yet all this time

Bijleveld, Nikolaj, Arjen Dijkstra, and Samuel Gessner, eds. *Cultivating and Communicating Natural and Technical Knowledge around 1800: Devotees of Science.*
Taylor & Francis Group, 2025
DOI 10.5117/9789463728577_CH02

Illustration 1. Jean Henri van Swinden (1746–1823). Oil on canvas, anonymous, early 19th century (*Collection of the University Library of the University of Amsterdam, 000.095*)

Van Swinden had been ignorant of the planetarium under construction in his very place of residence. On his first visit (with two colleagues from the university) Van Swinden was stupefied and immediately set himself up as the ambassador of Eisinga and his planetarium in the learned world. He wrote letters to colleagues in Paris and had them published in journals all over Europe (and North America). He also wrote a small booklet (in Dutch) on the planetarium, *Beschryving van een konststuk verbeeldende een volledig beweeglyk hemelgestel uitgedagd en vervaardigd door Eise Eisinga* (Description of a piece of art representing a complete and running planetarium devised and built by Eise Eisinga). This booklet was favourably reviewed in several other journals in Europe and added to Eisinga's instant celebrity, resulting in a constant stream of visitors to his house and attempts to incorporate the planetarium in the University of Franeker.[1]

In his publications, Van Swinden introduced Eisinga as a lonely 'genius' and ever since, Eise Eisinga has been presented as an exceptional figure in the late eighteenth-century world of science and learning, an artisan or a small business man without real astronomical education who nevertheless succeeded in building a highly complicated planetarium that still amazes the world. The odd fact that he installed his planetarium in the ceiling of his cramped living room at home, in the small city of Franeker, added to this picture of a somewhat exotic 'man of science'. Recent research, especially by Arjen Dijkstra, has shown that this picture is no longer adequate – much of it is the result of Van Swinden's framing of Eisinga as a lone genius and the romantic imagination of other writers about Eisinga. At the same time, we still do not have a clear and positive idea of how to locate Eisinga in the knowledge culture of the late eighteenth century.[2] The key to understanding the phenomenon of Eise Eisinga – and others like him – is the fact that Eisinga was in a sense discovered and instrumentalised by a university professor with extensive contacts in the world of the universities and academies of science of the late eighteenth century.

The So-Called Decline of the Dutch Universities

By the time Eise Eisinga decided to build a planetarium, the traditional world of science and learning in Europe was in disarray. Universities had long been the central element in the knowledge infrastructure and although the history of the institution went back more than five centuries, it was still flourishing (or again flourishing) in the seventeenth and even in the early eighteenth centuries. During the second half of the eighteenth century, however, criticism of the university system had become widespread and new institutions and educational opportunities had manifested themselves. Whether one speaks of decay, like many contemporaries did, or of transformation, like modern historians tend to do, the situation became rather chaotic. It was only during the revolutionary period at the end of the

1 Dijkstra, *Hemelbouwer*, pp. 87–94. On Van Swinden, who in 1785 left the University of Franeker and moved to the Athenaeum illustre in Amsterdam: Van Hoorn, 'De "gemeenebestgezindheid"', pp. 227–53; Van Helden, 'Jan Hendrik van Swinden', pp. 574–76.

2 Dijkstra, *Hemelbouwer*.

Illustration 2. The University of Franeker in 1786. Engraving by Dirk de Jong (*Collection Rijksmuseum Amsterdam, RP-P-1878-A-2128*)

century that attempts were made to systematically reconstruct the world of science and learning in Europe.

The example of Eisinga's Dutch Republic, a new country that had only emerged in the late sixteenth century, nicely illustrates the ups and downs of the history of the universities.[3] Before the Dutch Revolt, which started in the 1560s, the only university in the Netherlands was located in the Flemish city of Leuven, near Brussels, founded in 1425.[4] By the time the independence of the Dutch Republic was recognised by the Spanish king, the former sovereign of all the Low Countries, in 1648, the new state, or actually the provinces that constituted the new entity we call the Dutch Republic, had founded five new universities, located in the cities of Leiden

3 The most recent introduction (in English) to the history of the Dutch universities in the early modern period is: Van Berkel et al., *A History of Science in the Netherlands*. Since then, however, some of the still existing universities (Leiden, Groningen, Amsterdam) have commissioned multivolume histories that offer many new perspectives. For Franeker see: Jensma et al., *Universiteit te Franeker 1585–1811*.

4 Strictly speaking, this is not correct. In 1559 the Spanish king, Philip II, established a second university in the Netherlands, in the city of Douai, in the French-speaking region of Artois and near the city of Lille. But lectures only started in 1562 and although the first chancellor was a professor from Oxford, the main body of the professors came from Leuven. It is fair to say that during the first decades of its existence, the University of Douai was just an outpost of Leuven University.

(1575), Franeker (1585), Groningen (1614), Utrecht (1636), and Harderwijk (1648) (Illustration 2). Furthermore, quite a few cities, including Amsterdam, founded so-called 'illustrious schools', which provided basic academic education without having the right to grant a doctorate to its students. The *ius promovendi* remained the privilege of the universities. Teaching was done in Latin, which was the norm elsewhere in Europe.

All of these universities blossomed at least until the last quarter of the seventeenth century. They attracted hundreds of students, also from abroad, and many of the luminaries of seventeenth-century science and learning were employed as professors at one of the universities or illustrious schools. The fact that the international Republic of Letters (the network of the learned in Europe, both inside and outside of the university) was kept together by the use of Latin in letters and publications, very much facilitated the international exchange of students and professors. It is of course true that some of the most innovative scholars and scientists of the seventeenth century were not employed by the university, such as Christiaan Huygens (1629–1795) in Holland and René Descartes (1596–1650) in France (and the Netherlands), but they all had some kind of academic training; the university and colleges were still the gateway to a life in pursuit of knowledge.

From the late seventeenth century onwards, however, criticism of the university system grew stronger. Criticism had never been completely absent, as the writings of Paracelsians on the Continent and the Puritans in England testify. These authors complained about the purely theoretical nature of academic learning, the lack of space for experiment and practical training, and the strict adherence to Latin as the language of teaching, which excluded people without classical education. At some universities the administration and the professors tried to incorporate new directions of research and new didactical ideals into the existing system. At Leiden University a separate institution for the training of surveyors and military engineers was established in 1600, following the University of Franeker, where in 1598 the professor of mathematics, Adriaan Metius (1571–1635), had been given permission to teach a course in the Dutch language especially for future surveyors and engineers. Until the end of the eighteenth century these courses in the vernacular were continued, not only at Franeker and Leiden, but also at Groningen and Utrecht. The status of the teachers of applied mathematics was rather low, but they were not excluded from the academic community.[5] Even more important was the introduction, after 1675, of physical and chemical cabinets and laboratories. This was in line

5 For the example of Franeker see: Dijkstra, *Between Academics*.

with the rise of experimental science, but the laboratories were only meant to demonstrate the tenets of physics or chemistry, not to train the students in doing modern experimental research.

These modest innovations were unable to counter the steady decline in student numbers at Dutch universities from the late seventeenth century onwards. In part this was the result of the return of more peaceful conditions in large parts of Europe, especially Germany. The outbreak of the Thirty Years War, which lasted from 1618 to 1648, had forced many students and professors to seek refuge in the neighbouring Dutch Republic with its many universities. That students in Germany were now able to get a decent academic education at home (and were stimulated by their princes to do so) had nothing to do with the quality, or the lack of it, of the Dutch academies. Until deep in the eighteenth century, Dutch universities were able to attract first-rate scholars and scientists, people such as Herman Boerhaave (1668–1738) at Leiden and Petrus Camper (1722–1789) at Franeker, Amsterdam, and Groningen.

On the other hand, the career of Petrus Camper makes it clear that there were certain restrictions to enlightened reform at Dutch universities. In 1763 he was lured from Amsterdam to Groningen and was promised full support for his innovative research in anatomy and surgery. Camper took full advantage of the support he received from the administrators, but when in 1772 his colleague Frederik Adolf van der Marck (1719–1800), who was a fierce proponent of the introduction of public law, was rumoured to question the theological dogmas that all professors had to ratify on their nomination, Camper too ran into problems. Privately he was sympathetic towards Van der Marck's critical ideas, but publicly he felt he had to support the authorities even if they wanted to dismiss and expel Van der Marck. In the end, Van der Marck was indeed dismissed (with the endorsement of Camper), but already a year later Camper felt it would be safer for him also to leave Groningen. He retired to his estate near Franeker and until his death in 1789 remained an independent scientist.[6]

Given the unmistakable merits of people such as Boerhaave and Camper, framing the undeniable contraction of the Dutch universities in the late seventeenth and eighteenth centuries simply in terms of decline and decay is questionable. What is more important is the loss of function or even the growing irrelevance of academic education.

In the Dutch Republic, this loss of function of traditional types of education was visible not only at the university level, but also at the level of

6 Van Berkel, 'Petrus Camper', pp. 75–90.

secondary education. Traditionally, the Latin School, with its heavy stress on the classical languages, provided the necessary preparation for the university. Since Latin was the *lingua franca* of academic life, a decent training in the classical languages remained important. The level of proficiency may not always have been satisfactory, but to understand a professor or to write a short dissertation, a basic knowledge of Latin grammar and vocabulary was indispensable. Strictly speaking, a diploma from the Latin School was not required for entering the university; one could also acquire the necessary skills through private education or with the help of a local minister of the church. Yet the steady decline in the number of pupils at Latin Schools in Holland and other provinces indicates that fewer and fewer people were interested in acquiring a classical education. At the same time, an expanding network of new types of schools and private lecturers offered alternative opportunities for the young and the ambitious to get a decent, up-to-date education.[7]

Scientific Societies and the Quest for Useful Knowledge

Behind this shift in the orientation of education was a new desire for knowledge that was useful. Not only in the sense that knowledge could be applied, but also in the sense that it included more adequate professional qualifications. Traditionally, the Latin School and the universities had valued the contemplative life more than the practical or active life. At the university, theory had always been more important than practice. Even studies that were meant to be training for a specific profession, such as lawyer or physician, hardly prepared the student for the practical work a lawyer or a physician had to do. In the Faculty of Law, many hours were spent on Roman law, which was no longer valid in the Dutch provinces, while the provincial or national laws that were in force were not touched upon. Once a student had become a lawyer and left the university, he still had to get an attorney, a non-academic profession, to teach him the rules and conventions that applied in the court room. In the Faculty of Medicine, the medical doctor may have been well-educated in physiology and anatomy, but he had not learned to perform a decent surgery. This had to be done by a surgeon, another non-academic profession. This state of affairs was under attack during most of the eighteenth century, when progressive professors

7 For a critical reassessment of the decline thesis see: Frijhoff, 'Crisis of modernisering?', pp. 37–56.

tried to introduce their students to the practices of their profession. Van der Marck at Groningen introduced the study of common law in the university (to the detriment of Roman law) and his colleague Camper integrated surgery and anatomy into his courses. But they were still the exceptions to the rule that at the university teaching was predominantly theoretical and not practical.

During the second half of the eighteenth century, the quest for useful knowledge was the driving force behind the rise of a new kind of institution, the scientific and learned society. A national academy of science, like the Royal Society in England and the Académie des Sciences in France, did not exist in the Dutch Republic. In this highly federalised country, with a weak prince as its head (officially the Prince of Orange, also called the stadtholder, was only the servant of the provincial states), circumstances for a central academy of science were unfavourable. It was only at the provincial level that regents got together to establish a society that was to promote science and apply it for the benefit of the country. In most cases the regents took the initiative, and also established themselves as directors of the society. But the scientific work was done by its 'working members', usually university-educated people from all over the country. They drafted prize competitions, judged the submitted papers, edited the publications, and acted as lecturers when a society organised lectures and demonstrations. In this way, the internal organisation of the society mirrored and reinforced the social divisions that existed in society at large. Membership of a learned society was considered to be an honour and people from the lower strata of society (including attorneys, surgeons, and instrument makers) were as a rule denied membership.

The new learned societies emerged as an alternative focal point of science and learning in the second half of the eighteenth century. Universities were not always pleased with these new scientific institutions. Leiden University actively tried to block the establishment or to curtail the range of activities of the Holland Society of Sciences that was founded in 1752 in the province of Holland (in the city of Haarlem). At the same time, Leiden professors who in their academic courses hardly dealt with practical issues, were very active in the Holland Society in initiating research that would solve practical problems that cities and provinces faced on a daily basis (especially with regard to water management). Elsewhere, in provinces where the university and the learned society were located in the same city, usually the capital of the province, the university and the provincial society cooperated to their mutual benefit and actually formed some sort of symbiosis. In Groningen, law professor Van der Marck used the local society, Pro Excolendo Jure

Patriae, which in 1761 he had founded himself, as a vehicle for promoting the study of provincial and common law at the university. This cooperation can also be found in Utrecht, but not in the province of Friesland, where the university (of Franeker) had not been established in the provincial capital (Leeuwarden).

In a province, there was only room for one or two major scientific societies. By applying for a privilege and the nomination of the stadtholder as its 'protector', these societies tried to establish a monopoly in their respective provinces. Yet in the smaller cities the local elite organised its own local society for the promotion of science and scholarship. These local societies usually did not issue prize competitions, nor did they publish proceedings or 'memoirs', but they did organise courses given by travelling lecturers for the entertainment and enlightenment of its members (and sometimes the spouses and daughters of the members). To a greater degree than the larger societies, these smaller societies also aimed to educate its own members and to popularise current science among the educated public. They also met the growing demand for social interaction among the (local) elite. In the eighteenth century, the new civic ideal of sociability – the idea that people become better human beings when they work together on the same issues and learn from each other – also gained ground in the Dutch Republic and the new societies were a reflection of this ideal. Like the larger societies, the smaller societies maintained and complied with the existing social stratification, cultivating equality among their members, but keeping out those who were not considered to be of equal rank. The conservative nature of the societies also reflects the fact that one of the uses of science they envisioned was its support of the social and religious order. Science, if rightly interpreted, confirmed the existence of a wise and almighty God, who had established the world according to rules that humankind should not question. Useful knowledge thus was not just technical and medical knowledge, but also knowledge about the religious implications of scientific research.

The European University System in Times of Change

Many of the characteristics of the shifting intellectual landscape of the Dutch Republic in the eighteenth century are evident in other European countries. Complaints about the university system that were heard in the Netherlands were also to be heard elsewhere: too many universities for too few students, holding on to Latin as the language of instruction, no orientation to the needs of society, and not enough room for modern sciences and sciences that offer practical training. Each country had its own characteristics, such as the virtuoso tradition in England and the influence of the religious orders in countries such as Spain, Italy, and France. But everywhere universities were criticised, partial reforms were undertaken (as, for example, in Rome, or Coimbra), and learned societies sprouted across the continent.[8]

Although universities have always been subject to criticism, in the eighteenth century it became more general and more fundamental than ever. Universities represented the 'old order' as much as the church did, and they seemed destined for demise as much as the 'old order' did. Universities were charged with being unable to integrate new disciplines into their system of faculties (for instance, physics and mathematics were taught in the philosophical faculty, chemistry and botany in the medical faculty); didactics, with its disputations and theses and its heavy stress on oral presentations and rhetorical skills, was outmoded and unfit for research; retaining Latin as the language of instruction went against the rising tide of the vernacular; enforced religious adherence of at least the professors and the dominance of the theological faculty contradicted the new mood of religious toleration; the eroded rituals were no longer understood; and the dominance of the old elite irritated the rising middle classes. The universities survived because they had a monopoly on conferring degrees on lawyers and physicians, and in addition because a doctorate still counted as a symbol of social status for high-ranking administrators and regents. Contemporaries wanted the universities to become engines of progress, but they concluded that the universities were in fact symbols or even strongholds of traditionalism.

In contrast to the Dutch Republic, where a deeply entrenched provincial particularism thwarted every attempt to reform the system, elsewhere in Europe attempts were made to introduce some new elements to the university system, either by founding new universities or by adapting existing universities to modern circumstances.[9] An early example is the foundation of the new University of Halle in Prussia in 1694, an institution that was meant to be an alternative and counterweight to the old and archaic University of Wittenberg in Saxony, once the home of Martin Luther and Philip Melanchton and a centre of Lutheran dogmatism. Halle embraced freedom of teaching, *libertas philosophandi*, and German was introduced as the language of instruction. With the philosopher Christian Wolff (1679–1754),

8 For a general account of the history of the European universities in the early modern period see: De Ridder-Symoens, *A History of the University in Europe*. See also: Brockliss, 'Science', pp. 44–86.

9 For the following see: Frijhoff, 'Patterns', pp. 43–110; Frijhoff, 'Verlichte kaalslag', pp. 23–45.

who lectured at Halle from 1706 to 1723 and again from 1740 to his death in 1754, the new university became the centre of philosophical rationalism and the German Enlightenment. The conflict that led to Wolff's downfall in 1723 (he was accused of atheism and fatalism by the Pietists) has become known as one of the most important battles between religious dogmatism and enlightened rationalism in the eighteenth century. His triumphant reinstatement in 1740 as professor at Halle was a major victory for the forces of renewal and academic freedom. Another innovation was the introduction of the so-called *Kameralwissenschaften*, the science of public administration that aimed at stronger management of a centralised economy mainly for the benefit of the state. In 1727, Halle and Frankfurt an der Oder, another Prussian university, were the first to get a chair in Cameralism.

In 1734, following the example of Halle, George II, King of Britain and Elector of Hanover, established a new university at Göttingen (Lower Saxony), explicitly intending that it would be a centre for the ideas associated with the European Enlightenment. As in Halle, there was freedom of teaching, and the Faculty of Theology no longer had a say on teaching in other faculties. Professors were explicitly ordered to conduct and publish research as part of their assignment. Another innovation was the establishment of certain centres of research that explicitly focused on specific aspects of society, such as Johann Friedrich Blumenbach's anthropological museum. The university became especially known for its Faculty of Law, where Johann Stephan Pütter, a pupil of Wolff at Halle, taught public law (*ius publicum*) from 1746 to his death in 1807. The new university also became famous for its school of history, where historians combined the critical methods of their humanistic predecessors with the methods of philosophical historians such as Voltaire and Gibbon. The king and his prime minister in Hanover, Gerlach Adolph von Münchhausen, wanted the university to be no longer a secluded space for scholars and students, but an institution open to the needs of society.

The creation of new institutions was one way to reform the university system, the abolishment of small and antiquated universities another. The latter happened mostly in absolutist states in southern Europe. For instance, in 1714 in Spain, seven small Catalonian universities were closed after the War of the Spanish Succession. Instead, the Spanish king, Philip V, established a new university at Cervera, which was led by the Jesuit order. In Portugal the Marquis de Pombal ordered the closing down of the Jesuit-led University of Évora and the centralisation of academic teaching at the drastically reformed University of Coimbra. In France, universities in small towns were transplanted to the capital of the province, from Dole to Besançon (1691), from Molsheim to Strassbourg (1701) and from Pont-à-Mousson to Nancy

(1768); in 1751 the University of Cahors was disbanded and unified with the University of Toulouse. In these cases, the 'reform' was not meant to alter the system as such – that is, to orient it much more than in the past to the needs of society and to introduce new teaching methods like experimental classes – but to strengthen the hold of the government over the sometimes rebellious smaller universities outside the centres of power, or to break the power of the Church (and its religious orders). As in Germany, with the introduction of the *Kameralwissenschaften*, reform was primarily meant to strengthen the position of the state.

It was only after the French Revolution (1789) and its extension to other European countries that the combined effort of demolishing the old system and introducing new types of education, including a reformed university system, was speeded up. In 1793 all universities in France were dismantled and replaced by a few professional schools in the provincial capitals and in Paris: medical schools and law schools, a teacher training institution (the École Normale), academies for pharmacy and surgery, plus a few technical schools, such as the École Polytechnique and the École des Ponts et Chaussées.

After France started to export its revolution to other countries, one of the measures taken by the new regime was the demolition of many of the old universities in the occupied territories. One of the first 'victims' was the ancient University of Leuven. Already in 1788–1790 the Austrian emperor, Joseph II, who also ruled the Southern (former Spanish) Netherlands, had moved the Faculties of Law, Philosophy, and Medicine to Brussels, and now in 1797 the French closed down the remaining Faculty of Theology in Leuven. A little later, the wave of university extinction ('das grosse Universitätssterben') also swept through Germany. The renowned University of Cologne was shut down in 1798, and small universities such as Fulda, Rinteln, Helmstedt, and Altdorf closed their doors definitively (the University of Cologne was re-opened in 1919). *Gymnasia academica* or illustrious schools in the western part of Germany, such as those in Hamm, Meurs, Düsseldorf, Emmerich, and Hanau, suffered the same fate. While they may once have served a purpose (preparing students for continued study at one of the major universities), in the eighteenth century they only survived because of the wish of the local prince to have his own institution for higher education. Likewise, in the Dutch provinces, integrated into the French empire in 1810, all universities, including the one in Eise Eisinga's Franeker, were closed down in 1811, to be replaced by two clusters of independent faculties in Leiden and Groningen only (all under the umbrella of the newly created Imperial University in Paris). Already after the forced removal of the Prince of Orange (1795) and the creation of a unified political structure (the Batavian Republic), plans had been developed to merge all existing universities into one national university, but provincial particularism had prevented the execution of these policies until Napoleon forced the Dutch to comply with his designs for centralisation. It was only after the defeat of the French in 1813 that three of the old universities – Leiden, Groningen, and Utrecht – resurfaced, but they were soon reconstructed along modern lines.

However, the weeding out of existing and struggling universities went hand in hand with the creation of a range of new institutions. The Imperial University established by Napoleon in 1808 has already been mentioned. But this was not a university at all. It was a centralised organisation that supervised not only all institutions of higher learning in France, but also all primary and secondary education. The former individual universities (one in the territory of each Court of Appeal) were now known as academies, but still retained a rector and a local academic board. Most important was the new University of Berlin (1810), which was structured according to the ideas of the Prussian minister, scholar, and philosopher Wilhelm von Humboldt (1767–1835), who made personal *Bildung* the central goal of academic education and integrated teaching and research. Unlike the French Imperial University, the Humboldt University of Berlin granted full academic freedom to its professors. Whereas the highly centralised Imperial University in France eventually proved to be an obstacle to scientific progress (the institution was discontinued in 1896, supplanted by autonomous universities), the University of Berlin offered a model of modern academic teaching that was followed, partly at least, elsewhere in Germany and on the Continent (in England, nothing changed).

Scientific Societies and Academies

In the seventeenth century, growing discontent with the universities as centres of learning went hand in hand with the rise of a completely different kind of institution for scientific research, the scientific society and especially the academy of science.[10] Following a model provided by the informal gatherings of scholars in Renaissance Italy, early academies of

10 There is no general account of the history of the early modern academies of science comparable to the volume on the universities in Europe in that period edited by De Ridder-Symoens, *A History of the University in Europe*. A good place to start, however, is: McClellan, 'Scientific Institutions', pp. 87–106. For more details, see: McClellan, *Science Reorganized*.

history, arts and literature, and informal scholarly correspondence networks, scholars and scientists in several countries north of the Alps formed their own academies of natural science, now officially chartered.[11] The Royal Society of London for the Improvement of Natural Knowledge, founded in 1660 and chartered in 1662, is usually regarded as the oldest scientific society, to be followed in 1666 by the Académie royale des sciences (in English known as the Academy of Sciences) in Paris. The differences between these two societies were substantial. The Royal Society emerged from the activities of a group of scientists and scholars that was then officially chartered by the English king, but the king did not support the society financially and he, or his government, had no say in its day to day business. The Parisian Academy of Sciences, on the other hand, was founded on the initiative of finance minister J.-B. Colbert by King Louis XIV and was funded by the French state. Whereas the Royal Society included both scientists and wealthy amateurs and protectors of science (nobles and clergy whose donations also financed the society), membership of the Académie des Sciences was restricted to active scientists only. Both societies, however, enlisted members from abroad, such as the Dutch microscopist Antoni van Leeuwenhoek (1632–1723), who disseminated his findings through letters to the Royal Society, and the mathematician Christiaan Huygens, another Dutchman, who for a couple of years enjoyed an annual pension and a place of residence in Paris.[12]

Over the years, and especially after 1700, new academies were founded by princes all over Europe. Consequently, the eighteenth century has been rightly called 'the Age of the Academies'. These new academies had slightly different structures, varying numbers of members and corresponding members, objectives, and financial means. In 1677 the German emperor, Leopold I, promoted the already existing private Academia Naturae Curiosorum in Halle to the official position of Deutsche Akademie der Wissenschaften Leopoldina (the academy thus became an Imperial Academy in 1687). Another major addition was the foundation of an academy in Berlin, then the capital of Brandenburg-Prussia. In 1700, on the advice of the philosopher,

11 In Italy, both the papal and the national academy of sciences claim their descent from the Accademia dei Lincei, founded in 1603 by the Roman aristocrat Federico Cesi and best known for Galileo's membership. In reality, the Accademia dei Lincei was dissolved in the 1630s and the two national academies date from the nineteenth century. In the meantime, Italy did count other academies, like the Accademia del Cimento (Academy of Experiment) in Florence (1657–1667).

12 McClellan, 'Scientific Institutions', p. 92, distinguishes academies (tightly structured, state funded, with restricted membership, and with official duties) from societies (loosely structured, privately funded, larger membership, no official duties). At the same time, he acknowledges that the actual name of the institution does not always indicate the type of institution it belonged to.

mathematician, and historian Gottfried Wilhelm Leibniz (1646–1716), the Elector of Brandenburg, Frederick III, established the Kurfürstliche Brandenburgische Societät der Wissenschaften (Electoral Brandenburg Society of Sciences), which, after Frederick III became king of Prussia (1701), changed its name in Königliche Preussische Sozietät der Wissenschaften. Unlike the Parisian Academy of Sciences, this new institution was not financed by the state, but earned its income from the monopoly of production and sale of calendars in Prussia. And unlike both the London Royal Society and the Paris Academy of Sciences, the Prussian Academy was intended to further both the natural sciences and the humanities. For this purpose, the Academy was divided into at first four, later two divisions, one for the natural sciences and one for the humanities. It thereby set an example that was followed by other German academies, such as the Bavarian Academy of Sciences (1759), with two divisions or classes, one for philosophy (including mathematics and the natural sciences) and one for history (here as elsewhere one should be aware of the fact that the German word 'Wissenschaften' refers to both the natural sciences and the humanities). In other countries, sometimes separate academies were founded for either the natural sciences or the humanities. In Sweden the Royal Swedish Academy of Sciences was founded in Stockholm in 1739 as a counterweight to the Royal Society of Sciences in the university town of Uppsala (1719), that still used Latin (and did so until 1863). In the eighteenth century other national academies of science were founded in St. Petersburg (1724), Hanover (1751), the Austrian Netherlands (1772), and Portugal (1779).

On a somewhat lower lever, cities all over Europe boasted many strictly private academies or societies, such as the scientific societies in Rome, Bordeaux, Toulouse, Nancy, and Haarlem. Some of these societies published *memoires* and proceedings, others did not; some appointed members from all over Europe, others restricted membership to local or national scholars and scientists; some focused on just the (natural) sciences, others also covered the arts and *belles lettres*. Membership of these societies was coveted greatly by scholars and scientists all over Europe, and the number of societies of which one was a member served as an indication of one's importance within the learned community. It was not uncommon for scholars to advertise their membership of an academy or society on the title page of their publications.

Academies and societies fostered the sciences in many ways. Members presented their findings in the weekly or monthly meetings of the societies and published them in the quarterly or yearly transactions or memoirs of the societies. Societies also used their funds for prize competitions for specifically set research questions. They awarded gold and sometimes silver medals for the best answers and published the prize-winning essays in their proceedings. Since prize competitions were not restricted to society members, the institutions in this way presented themselves as the arbiters of science, defining not only the most urgent topics for scientists but also awarding status to those who had won a competition. In some cases, academies furthermore discussed patent applications and advised the government to either grant or reject an application. Finally, the more prominent societies and academies undertook or sponsored scientific expeditions. Both the Royal Society and the Paris Academy dispatched astronomers and geographers to various places around the world to observe the transits of Venus in 1761 and 1769. The only way in which societies did not promote science was education. Universities all over Europe were very keen to keep the teaching of science to themselves.[13] When in 1752 the Holland Society of Science in Haarlem was founded, Leiden University, as mentioned before, immediately intervened and ensured that the society's charter explicitly excluded teaching as one of its activities. Instruction of the general public was left to the travelling lecturers and independent journals.

By the end of the eighteenth century the science landscape was dominated by two complementary, non-competing types of institution, the old (and sometimes partially reformed) university and the new scientific academy or society. To be a person of significance in science meant being affiliated to at least one of these institutions, either as a professor or as an *académicien*; many well-known scientists were both. A career in science outside these structures was hardly possible anymore. Individual princely patronage, as Galileo had sought and gained in the early seventeenth century in Tuscany, had almost disappeared in the eighteenth century. If a prince wanted to favour a researcher, he arranged an appointment as member of 'his' academy, as was the case with Pierre Louis Maupertuis, a prominent member of the Paris Academy of Science who in 1744 came to Berlin on the invitation of Frederick the Great and in 1746 was appointed president of the Royal Prussian Academy.

Intermediaries, Go-Betweens, and Other Outsiders: Devotees of Science

In the eighteenth century, the universities, academies, and scientific societies were undoubtedly the centres of science, the places where the most advanced science was practised and where authority and trustworthiness

13 An exception is the Russian Academy of Science in St. Petersburg, which was founded in 1724 as both a regular academy and a university.

were established. But this world of science was not isolated from society, far from it. First, the usefulness of science was one of the main articles of faith of eighteenth-century scientists and even the smallest scientific society was eager to contribute to the common good. Prize competitions aimed to solve specific technical questions or to strengthen the belief in an almighty and providential God. National observatories, often linked to or manned by members of an academy, such as the Royal Greenwich Observatory and the Royal Observatory in Paris, produced maps, calendars, nautical works, and other publications of obvious utility.

Furthermore, the world of official science could not have existed without the contributions of a multitude of people, often from the lower and middle classes, who devoted much of their life to gathering and disseminating knowledge but who typically had no university training and were excluded, for social as well as intellectual reasons, from the academies and scientific societies.[14] It is interesting to note that quite often these intermediaries belonged to religious orders or were ministers of the church, which gave them ample time to devote their lives to the dissemination or popularisation of science. Among these devotees, intermediaries, and mediators of science were, first of all, editors of journals in which scientists could publish the results of their research or in which the notes of the meetings of an academy or even the results of research, discussed in the academy but not yet published in its memoires, could be published. Abbé François Rozier (1734–1793), whose monthly *Observations sur la physique, sur l'histoire naturelle et sur les arts* (1772) contained news from the Paris Academy of Sciences and other institutions, contributed greatly to the speedy circulation of knowledge in late eighteenth-century Europe.[15] Then we have collectors of unknown botanical specimens in the colonies, translators who facilitated the circulation of knowledge now that Latin had lost its monopoly on science communication, librarians who collected and distributed established science, artists who drew the pictures that more than ever were instrumental in communicating new findings, and skilled technicians and instrument makers who provided scientists with the new devices needed for scientific research.

Most of these outsiders were men, but many audacious women also contributed to science. Their contributions were recognised in their own days

14 See: Schaffer et al., *The Brokered World*. See also: McOuat and Stewart, *Spaces of Enlightenment Science*. Different languages have different words to denote outsiders with scientific interest and expertise, but the word 'amateur' is now completely discredited. Often researchers take refuge in contemporary terms, such as 'artistes' in French and 'konstgenoten' in Dutch, but these terms do not always refer to the same group.

15 McClellan, 'Scientific Institutions', p. 96.

and were certainly not overlooked by contemporaries, although these women were later written out of history. In 1787 the French chemist A.F. Fourcroy even wrote, optimistically: 'It is luckily recognised for the common good that [women] are capable of the efforts necessary to successfully cultivate the sciences'.[16] To name just one example, the French naturalist Madeleine Françoise Basseporte (1701–1780) was the protégé of Bernard de Jussieu and Comte de Buffon and devoted much of her life to dissecting and illustrating plants in the Jardin du Roi in Paris and elsewhere. She was the first woman to be granted a salaried position at the Jardin; beginning in 1741, she served for forty years in the position of 'Peintre du Roy', to the full satisfaction of her male superiors. Rousseau famously remarked about her: 'Nature gives plants their existence, but Mlle Basseporte preserves it for them'.[17]

These outsiders, male or female, often worked in places far removed from the hallowed halls of the universities or the exclusive meeting rooms of the academies and scientific societies. Still, they contributed significantly – directly and indirectly – to the expanding body of knowledge and the well-oiled machinery of the scientific movement. The state of flux in which the scientific institutions in the age of the Enlightenment found themselves, favoured these outsiders, or offered them several opportunities to make meaningful contributions to science.

Another category of non-academic but essential contributors to science consists of those who popularised science in the eighteenth century and spread the notion that science is not just a pleasant pastime, but also meaningful and useful. State bureaucracies created national academies because they believed in the usefulness of science, but the smaller societies were also exponents of this belief. Without this widespread popular belief in the utility of science and the popular support for science, the scientific movement in the eighteenth century could not have flourished in the way it actually did. The popularisers of science, although not contributing to science as such, indirectly supported the efforts of the academics and the academicians. This group included not only the itinerant lecturers, who offered courses in chemistry or physics to the general public, or the members of the small local societies, but also the hack writers and petty journalists who ignited the popular imagination with stories about new inventions or sensational cures.

What is important to note, however, is that in order to make a lasting imprint on science, these non-academic devotees to science had to have some connection to the institutions that defined science – that is, the

16 As quoted in: Gelbart, *Minerva's French Sisters*, p. 260.

17 As quoted in: Gelbart, *Minerva's French Sisters*, p. 108.

universities and the societies or academies. In a few cases these devotees of science were able to enter the official world of science themselves. Antoni van Leeuwenhoek is a classic example. In the 1670s this draper from the city of Delft in Holland, whose curiosity for things hitherto unseen was insatiable, manufactured single-lens microscopes that were unmatched in Europe. Despite his modest social status and his lack of language skills (his knowledge of Latin was rudimentary at most and his letters to the London Royal Society were written in Dutch) he was elected a member of the Royal Society in 1680.[18] His contemporary Daniel Gabriel Fahrenheit (1686–1736) offers another example. This German merchant, travelling lecturer in chemistry and glassblower living in the Dutch Republic, who was famous for his highly reliable thermometers, was admitted to the Royal Society in 1724. The majority of the non-academic devotees of science was not admitted, but they too needed some sort of connection to a society to ensure that their efforts would gain the attention of the scientific community.

It is rarely the case that outsiders organised themselves. A famous example is the French Société des Arts, founded in 1732 in Paris to further the interests of the so-called *artistes*, architects, clockmakers, and other workmen who considered themselves superior to ordinary artisans and who claimed that their work, though not philosophical by nature, nevertheless required 'esprit' or intelligence.[19] This is more or less Diderot's definition of the *artistes* in his *Encyclopédie* article on 'Arts' (1751), which gave the workmen at least some kind of recognition. Diderot's acknowledgement, however, was somewhat ambivalent. He made it clear that at the end of the day, the practical nature of the work of the *artistes* disqualified them in the eyes of those whose work was essentially philosophical, the *savants*. He thus reaffirmed the predominance of theoretical work over practical activities, while at the same time allowing the more mindful artisans to claim their own place just below the *savants*.[20] They were excluded, however, by the Parisian Academy of Sciences and suffered from the Academy's disdain for artisanal work. The *artistes* thus gathered in their own Société and claimed that they too had an essential function in the movement towards moral and material improvement that we have dubbed 'the Enlightenment'.

So even in the case of the Société, the rule applied that outsiders needed some connection to the established centres of science in order to have an

18 Leeuwenhoek even acquired a reputation for his anti-intellectual stance. See: Van Berkel, 'Intellectuals against Leeuwenhoek', pp. 187–209.

19 Bertucci, *Science and the Mechanical Arts*.

20 Bertucci, *Science and the Mechanical Arts*, pp. 4–5.

impact on the development of science. This is also where Eise Eisinga fits into the larger picture of eighteenth-century science. He acquired his skills and expertise outside the academic sphere, but was nevertheless spotted by a university professor, Van Swinden, who acted as a gatekeeper for those who operated outside the universities and the academies. Van Swinden was truly fascinated by Eisinga and by his letters and publications; he turned him into a celebrity in the world of science. Van Swinden also advised the board of the university and the provincial government to incorporate the planetarium in the university, with Eisinga as its curator. Van Swinden's plan was to use the planetarium – or a copy made for the university – to modernise the teaching of astronomy at the ailing university. However, characteristically the conservative regents ignored the advice and only rewarded Eisinga, much to his chagrin, with a silver coffee pot and a silver teapot. The regents even added insult to injury by having the servant of the silversmith hand the presents over to Eisinga.[21] Small wonder that after the dissolution of the old regime in the Netherlands, Eisinga advised the new revolutionary government in the province of Friesland to fundamentally reorganise and modernise the University of Franeker – another project that failed, unfortunately.

Bibliography

Bertucci, P., *Science and the Mechanical Arts in Old Regime France* (New Haven 2017).

Brockliss, L., 'Science, the Universities, and Other Public Spaces: Teaching Science in Europe and the Americas', in: R. Porter, ed., *Eighteenth-Century Science* (Cambridge 2008), pp. 44–86.

De Ridder-Symoens, H., ed., *A History of the University in Europe*, vol. 2: *Universities in Early Modern Europe (1500–1800)* (Cambridge 1996).

Dijkstra, A., *Between Academics and Idiots: A Cultural History of Mathematics in the Dutch Province of Friesland (1600–1700)* [Doctoral thesis] (Enschede 2012).

Dijkstra, A., *De Hemelbouwer: Een biografie van Eise Eisinga* (Gorredijk 2021).

Frijhoff, W.Th., 'Crisis of modernisering? Hypothesen over de ontwikkeling van het voortgezet en hoger onderwijs in Holland in de 18e eeuw', *Holland* 17 (1985), pp. 37–56.

Frijhoff, W.Th., 'Patterns', in: De Ridder-Symoens, *History of the University in Europe*, vol. 2, pp. 43–110.

21 Dijkstra, *Hemelbouwer*, pp. 98–99.

Frijhoff, W.Th., 'Verlichte kaalslag in Europa? De academische herstructurering rond de Revolutietijd en haar gevolgen', *De Vrije Fries* 91 (2011), pp. 23–45.

Gelbart, N.R., *Minerva's French Sisters: Women of Science in Enlightenment France* (New Haven 2021).

Jensma, G.Th. et al., ed., *Universiteit te Franeker 1585–1811: Bijdragen tot de geschiedenis van de Friese hogeschool* (Leeuwarden 1985).

McClellan, J.E., *Science Reorganized: Scientific Societies in the Eighteenth Century* (New York 1985).

McClellan, J.E., 'Scientific Institutions and the Organization of Science', in: R. Porter, ed., *Eighteenth-Century Science* (Cambridge 2008), pp. 87–106.

McOuat, G., and L. Stewart, ed., *Spaces of Enlightenment Science* (Leiden 2022).

Schaffer, S. et al., ed., *The Brokered World: Go-Betweens and Global Intelligence, 1770–1820* (Sagamore Beach 2009).

Van Berkel, K., 'Intellectuals against Leeuwenhoek: Controversies about the Style and Methods of a Self-Taught Scientist', in: L.C. Palm and H.A.M. Snelders, ed., *Antoni van Leeuwenhoek (1632–1723)* (Amsterdam 1982), pp. 187–209.

Van Berkel, K., 'Petrus Camper and the Limits of the Enlightenment', in: K. van Berkel and B. Ramakers, ed., *Petrus Camper in Context: Science, the Arts, and Society in Eighteenth-Century Dutch Republic* (Hilversum 2015), pp. 75–90.

Van Berkel, K. et al., ed., *A History of Science in the Netherlands: Survey, Themes, and Reference* (Leiden 1999).

Van Helden, A., 'Jan Hendrik van Swinden 1746–1823', in: Van Berkel et al., *History of Science in the Netherlands*, pp. 574–76.

Van Hoorn, M., 'De "gemeenebestgezindheid" van Jan Hendrik van Swinden (1746–1823): Een schets van zijn politieke bedrijvigheid in de jaren 1795/1796 en 1800/1801', in: E.O.G. Haitsma Mulier et al., ed., *Athenaeum Illustre: Elf studies over de Amsterdamse Doorluchtige School, 1632–1877* (Amsterdam 1997), pp. 227–53.

About the Author

Klaas van Berkel is emeritus professor of modern history at the University of Groningen.

3. Eise Eisinga's World: The Profile of a Devotee to Astronomy between Enlightenment and Romanticism

Arjen Dijkstra

Abstract: Eise Eisinga, the builder of the Franeker Planetarium, has long been understood as a remarkable figure, whose perseverance gave him the skills to build one of the largest mechanical planetariums in the world. He has been depicted as an autodidact, working without anybody knowing of his work. This picture is not correct and prevents us from understanding him as a product of his time and circumstances. This chapter approaches Eisinga as a true devotee to knowledge and the study of astronomy and mathematics. It traces Eisinga's connections to people across the Netherlands and gives insight in how knowledge and mathematics were part of a world which existed on the doorstep of universities for the better part of the eighteenth century.

Keywords: planetarium, orrery, University of Franeker, Friesland

On 22 February 1780, Jean Henri van Swinden (1746–1823), professor of natural philosophy at the University of Franeker in the north of the Dutch Republic, visited the local master wool comber and merchant Eise Eisinga

This article is for a large part a reworking of a report I wrote for the Royal Eise Eisinga Planetarium. On the road leading to their nomination to become included on the UNESCO list of World Heritage I studied Eisinga's life and the history of the Planetarium. This resulted in the report 'Eise Eisinga en het Planetarium' (Groningen 2019), which was largely published as 'The Planetarium and Eise Eisinga' in: *The Royal Eise Eisinga Planetarium*, vol. 3, pp. 64–95. My original report was translated by Joke de Groot, Hawkeye Vertalingen en Redactie. I kindly thank her for letting me use her translation as the basis for this article.

Bijleveld, Nikolaj, Arjen Dijkstra, and Samuel Gessner, eds. *Cultivating and Communicating Natural and Technical Knowledge around 1800: Devotees of Science.*
Taylor & Francis Group, 2025
DOI 10.5117/9789463728577_CH03

(1744–1828). He was told that Eisinga had made a planetarium. Van Swinden expected to see a table model orrery, but that was not what he found. Instead he was struck with awe and amazement to find that Eisinga had built the largest ever planetarium, in the ceiling of his living room.[1] Within weeks of his first visit, Van Swinden wrote to a handful correspondents an account that was published in learned journals and newspapers across the world. And later that year, he published a complete description of the planetarium, which he sent to even more correspondents. The Franeker Planetarium and Eisinga were an instant success.[2]

Van Swinden and Eisinga would become key figures in the history of the natural sciences of the Netherlands. Van Swinden can be considered as one of the most important Dutch academics of his time involved in astronomy, meteorology, physics, and philosophy. Eisinga's planetarium would soon become one of the best visited astronomy sites of the Netherlands and with it Eisinga became a national celebrity. Of his own accord he had devoted his life to astronomy, there was no institution which supported him, there was no formal or even informal structure that expected him to spend so much of his time and money on astronomy.

Eise Eisinga has long been understood as a remarkable figure, with an extraordinary aptitude for mathematics. He had the perseverance and skills to build one of the largest mechanical planetariums the world has ever seen. However, in nineteenth- and twentieth-century accounts, Eisinga's humble origins and his solitude were often stressed, as well as the simplicity of his character. He is even said to have been an autodidact, working without anybody knowing of his work. The sources of Eisinga's time tell a different story, and with this paper, I will try to move past this romantic image.[3] What can we reveal about Eisinga's life if we return to the sources?

1 Halbertsma, *Hulde aan Gysb*ert, p. 305. On how Eisinga's planetarium compares to others see King and Millburn, *Geared to the stars*.

2 See Dijkstra, *Hemelbouwer*; Vreeling 'Eise Eisinga was hier'; Terpstra, *Friesche sterrekonst*, and for an English introduction Mulder de Ridder, 'Eise Eisinga'. I need to stress that both Terpstra and Mulder de Ridder have a very different view on Eisinga than I have. On the University of Franeker see Jensma, et al., *Universiteit te Franeker, 1585–1811* and Van Sluis, *De academie van Vriesland*; on mathematics and astronomy at that university see Dijkstra, *Between academics and idiots*.

3 In the autumn of 2024, a new Dutch biography of Eisinga was published: Langereis, *Machine-man*. This book stresses Eisenga's virtue, but relates that to the personal sphere and presents him primarily as a genius. Langereis promises to offer a new perspective on Eisinga but actually returns to some of the myths that have little historical basis. While the book does ponder extensively on the fragmentary personal sources on Eisinga's life, it regretfully obfuscates work done by other historians understanding those and fails to introduce new sources.

In reconstructing the image of Eisinga by tracing his life, career and his social activities, inevitably new questions arise. My question is why the successful entrepreneur and politically active Eisinga was also devoted to the study of astronomy and mathematics? I will argue that this is because he was brought up in a culture where studying these fields held value and prestige. This culture can be found in many parts of Europe and so too in the north of the Netherlands. Eisinga, therefore, needs to be understood as one of the many devotees of his time, some of whom worked in wide and loosely organised networks. Crucial for his success is that he profited from the nearby university and also found a way into academic circles, although he never became a professor or a full academic. But Eisinga was not really aiming for that, he was looking for recognition from a broader audience.

Wool Combing

Eise Eisinga was born on 21 February 1744 in Dronrijp, a village between the towns of Leeuwarden and Franeker. He came from a family of wool combers and merchants. His father Jelte Eises and mother Hitje Steffens owned a wool combery or cardery. His father and his uncle were independent master wool combers, just like their father had been.[4] Eise would follow in his father's footsteps, too and become a master wool comber. He also shared with his father a strong interest in technological devices (Illustration 3).

For long Eisinga was seen as a simple artisan and is sometimes even referred to as being of modest income and humble background.[5] This derives from a poor understanding of the wool trade, and of what a wool comber could be in Friesland. In general, 'wool combing' refers to exactly what the term would suggest: the combing of wool. In the early modern period (and well before that) wool was one of the most important raw materials for clothing.[6] Wool combing was basically low-skilled and poorly paid work, even lower paid than spinning, which was a next step in the production of wool for the textile market.[7] To put this into some perspective: spinning

4 All biographical data are based on Terpstra, *Friesche sterrekonst*, pp. 211–13 unless otherwise specified.

5 See for instance Havinga et al., *Planetariumboek*, pp. 19, 32; compare how he is frowned upon by Jacob van Lennep, see Mulder de Ridder, 'Eise Eisinga', pp. 81–82.

6 A fine description of the whole process is to be found in: Buys, *Nieuw en volkomen woordenboek*, vol. 10, pp. 850–52.

7 Van Nederveen Meerkerk, 'Market Wage', p. 169; Van Nederveen Meerkerk and Schmidt, 'Between Wage Labor', p. 723.

Illustration 3. Eise Eisinga (1744–1828). Oil painting by Willem Bartel van der Kooi, 1827 (*Collection Royal Eise Eisinga Planetarium Franeker*)

was sometimes done by prisoners as a form of forced labour. In some parts of Europe, therefore, wool combing was granted an even lower status than spinning. However, we should not draw a too literal understanding of Eisinga's profession.

In the province of Friesland, the term wool comber could have a broader meaning. Eisinga and his family were the owners of a business that dealt with virtually the entire processing of the wool after shearing. Everything from washing to colouring, spinning and even the production of fabrics and clothing found a place in the company. They were master wool combers, not poorly paid artisans.[8] An explanatory dictionary dating from the late eighteenth century described Friesland as a province where there,

> is no lack of factories of silk and woollen cloth [...]. A large number of wool combers are to be found there, who comb and prepare the wool, then do the spinning and give the sagathy to poor people for knitting stockings, gloves and bonnets, or send it to places outside the province.[9]

In the eighteenth century, Franeker was a centre of the wool industry. Around 1750, more than twenty master wool combers were active there, who together generated work for an estimated 2,000 people.[10] Next to dairy farming it was the most important source of income for the entire town. Various wool combers were seen as masters who directed and managed (undoubtedly often with their wives) middle-sized factories.

Eise Eisinga's family helped him set up one of these wool-combing businesses in Franeker. In 1768, his parents co-financed the purchase of the property 'De Ooijevaar' (The Stork), which was later to house the planetarium.[11] Besides the family investment, Eisinga took out a number of loans in order to be able to buy large quantities of wool.[12] Thus, Eisinga's wool combery in Franeker was very much like a branch of the family business: the risks were spread while the returns increased. Knowledge, materials, and tools were shared. Consequently, Eise Eisinga was able to get off to a flying start.[13] By the early 1770s, Eise Eisinga had settled as a master wool comber in Franeker; he now had time to spend on local politics, the church, and his other interests.

8 Dragt, 'Over wol', pp. 80–90.

9 'Een groot aantal wolkammers vind men 'er, die de wol kammen bereiden, vervolgens doen spinnen, en het sajet aan arme menschen geeven, om 'er kousen, handschoenen en mutzen van te breiden, of buiten de Provintie verzenden'. Kok, *Vaderlandsch woordenboek*, vol. 15-16, pp. 671.

10 Cf. Faber, *Drie eeuwen*, vol. 1, pp. 238–39; Faber, *Drie eeuwen*, vol. 2, p. 625; cf. also Nijboer, *'Het openen en sluiten'*, pp. 178–80.

11 Cf. Eisinga's letter to Van Swinden, 6 July 1793, UBL 755. The Planetarium's history of ownership was carefully researched by Gerard van der Heide, who generously shared this information with me.

12 See Tresoar, Nedergerecht Franeker (12-12), inv. no. 311 (courtesy to Martin Engels), cf. Eisinga to Scheltema, 6 September 1818, UVA.

13 A similar strategy was invariably used for his brothers and his sons. This did not always work.

Education in Mathematics and Astronomy

From the beginning of the twentieth century Eise Eisinga has been presented as an autodidact, a self-taught person who had become proficient without obvious help from a teacher and without having attended an educational institution.[14] This is a recurring qualification that does not, however, reflect the schooling Eisinga actually received. As a wool comber, Eisinga received thorough training in his parents' company. He was, after all, supposed to follow in his father's footsteps.[15] Judging by the results of his business, his training was a success. But he had enjoyed primary school as well and later received an education in mathematics and astronomy.[16] For the young, talented wool comber's son, the education was crucial, since it offered him the opportunity to look beyond the boundaries of his workshop and of his social standing.

His education started at the village school in Dronrijp.[17] Following that, from a young age he was taught mathematics and astronomy by Willem Wytses (d. 1773) in Franeker.[18] Wytses, who worked in the wool industry during the winters, had more students; it was a job on the side which earned him some extra income.[19] It is Wytses's teaching that ultimately resulted in a series of beautiful manuscripts by Eisinga, which have always found admirers. Eisinga composed the most impressive of these well before his twentieth birthday.[20] The manuscripts show how he developed from a student who makes elaborations of Euclid into someone who masters complicated astronomy.[21] What has gone unnoticed is that Wytses had likely based his lessons on a course he followed at the University of Franeker.[22] His teachings contained assignments from broadly available literature – the examples are

14 Sometimes he is even considered to be part of a group of autodidacts, see: Zuidervaart, 'De Friese "Boerenprofessor"'.

15 Van Swinden, *Beschryving* (1780), pp. 7–8.

16 On mathematics in primary schools in Friesland see Tresoar, Hs 629, esp. the note by J.H. Halbertsma.

17 Eekhoff, 'Het leven', p. 38.

18 Every week Eisinga was given a series of sums that he worked out and dated at home. This education took place from the summer of 1759 until the winter of 1760. The level of these lessons was in no way inferior to that of the mathematics being taught at Franeker University.

19 Cf. Eisinga to Scheltema, 6 September 1818, UVA.

20 Most of these have been digitised and are available at the Royal Eise Eisinga Planetarium in Franeker.

21 See also Van Swinden, *Beschryving* (1780), pp. 7–8.

22 See Tresoar Hs 627, which is likely a manuscript by Wytses and resembles the lessons Eisinga received to great detail, see Dijkstra, 'Spectaculaire vondst'.

taken from the likes of Willem Bartjens, who had published a standard Dutch mathematics book, and translated works of Benjamin Martin, which offered a more complex introduction to mathematics, physics, and astronomy – and his lessons closely followed the courses taught at university.

After his time with Wytses, Eisinga continued to improve his skills in the field of astronomy. He eventually worked out all the lunar and solar eclipses in Friesland between 1762 and 1800.[23] And he would compile and calculate a long catalogue with sundials.[24] Complicated arithmetical works showed the finesse with which Eisinga mastered these arts. The elaborate astronomical writings would circulate in Eisinga's social network. We know that his brother, for example, made copies of them. They have been found in the private collections of rich Frisian noble families as well as in church archives.[25] This shows how much he was a product of the academic culture of Franeker.

Eisinga was not the only who benefited, albeit indirectly, from the University of Franeker. There were many mathematics teachers in Friesland; even in Franeker there were 'arithmeticians' other than Wytses actively teaching boys in advanced arithmetic, bookkeeping, and astronomy. One of the best known of these mathematicians is without doubt Wytze Foppes (1707–1778), from the nearby village of Dongjum. Foppes would even become a courtier of the Frisian Nassaus; the stadtholder's family patronised him and his works.[26] At times he even made a public display of his status, for instance when he gathered an audience to observe the transit of Venus in 1761, witnessing him at work.[27] Apparently, being an astronomer had social value.

The mathematicians of eighteenth-century Frisia co-existed in a precarious balance: they profited from each other but were also competitors. While this tension has been observed on a number of occasions, it has never been

23 It is quite possible that he produced this manuscript without further teachers, but it is clear that he did it with the help of books and/or manuscripts. The language and structure are strongly reminiscent of and sometimes directly related to a number of publications dating from the course of the eighteenth century, see Dijkstra, *Hemelbouwer*, pp. 44–50.

24 Royal Eise Eisinga Planetarium, manuscript 17, see also Tresoar, Familiearchief Van Eysinga-Vegelin van Claerbergen (323), inv. no. 3538.

25 Tresoar, Familiearchief Van Eysinga-Vegelin van Claerbergen (323), inv. no. 3538 and Tresoar Hs 1131; Van Swinden, too, was immediately given access to this manuscript and showed himself particularly impressed by it. Van Swinden, *Beschryving* (1780), pp. 6–8.

26 See also: Ekama, *Oratio*; Terpstra, *Friesche sterrekonst*; Dijkstra, *Between Academics*, pp. 38–43; Eekhoff, 'Het leven'.

27 Eisinga is sometimes said to have assisted Foppes, this is, however, not the case. See Eisinga to Scheltema, 20 July 1818 and 17 August 1818, UVA. Cf. Wassenbergh, 'Friesche wiskundigen', pp. 83–84; Zuidervaart, *Speculatie*, p. 16.

described as a network that stimulated dissemination. Most studies do not advance further than a listing of Frisian mathematicians.[28] Nonetheless, being a remarkable mathematician or accomplishing an astronomical achievement could grant exceptional status in this world. Consequently, the Frisian mathematicians tried to outdo each other with instruments, students, and publications and were, to a certain extent, competitive.[29]

Mathematicians were not the only scholars active during the second half of the eighteenth century without being directly affiliated with universities. These other scholars studied, for example, the cattle plague, or even 'experimental philosophy'. They too built instruments, wrote pamphlets and brochures, and, briefly put, tried to understand the world and nature. This group includes the Frisian pastor Eelco Alta (1723–1798), who is always presented as an opponent of Eisinga, but also Yme Tigchelaar (1753–1812), whom we will meet again later, and Geert Reinders and Jacobus Uilkens from the province of Groningen.[30] It would seem contrived to draw too clear a dividing line between the mathematicians and the other devotees who worked outside of academia, particularly as they kept in close contact with one other. Eisinga generously shared his knowledge with the public and – albeit sparingly – commented on the work of other mathematicians.[31] Furthermore, this search for knowledge was certainly not limited to the north of the Netherlands, as we learn from the other contributions to this volume. All over Europe people were occupying themselves with a renewed understanding of nature, whether from a mathematical or an astronomical angle or not. Eisinga's development and work has to be understood as a specific case in this broader development.

Eise Eisinga's Social Position

When he settled in Franeker in 1768, it was immediately clear that Eisinga fostered ambitions. The building that housed his wool-combing business is

28 Zuidervaart gives an overview of these and makes one of the more integrated assessments of the tradition: Zuidervaart, 'De Friese "Boereprofessor"'; cf. Dijkstra, *Between Academics*, pp. 38–43.

29 Eisinga himself at times also tried to show how his skills outdid those of others, see Eisinga to Scheltema, 20 July 1818, UVA. For a broader view on how astronomy was practiced in the Dutch Republic at the time see Zuidervaart, *Van 'Konstgenoten' en hemelse fenomenen.*

30 See for example Tersteeg, *De levensschets*; Breuker, 'Acht maaie'; Breuker, *Eelko Alta*; Dijkstra, 'Het stokske'; Botke, *Boer en heer.* It would appear from this that Alta is Eisinga's counterpart rather than his major opponent.

31 Cf. Eisinga, *Vriesche koopman*, cf. Eisinga to Scheltema, 20 July 1818, UVA.

located right in the centre of Franeker, next to the townhall, a very prominent spot. He soon set out on a civil career, too: in the 1770s he became a member of the local guard, was a collector for the town poor; he next rose to become a member of the city's council, and in 1778 was promoted to guardian of the poor. He was clearly a man with aspirations in both his professional and public life. It may not have been uncommon, but he was remarkably successful.

As guardian of the poor Eisinga would present a 'Plan for the Establishment of a City Poor House'. He did so because he was asked by the city authorities to build a new poorhouse where 'old decrepit people, men and women, as well as orphaned children and children of poor widows in need' were to be given a place. In this plan he envisioned a house where order and peace reigned, where a 'bookkeeper' would be in charge, and where everyone had work to do. In addition, he proposed biweekly meetings between guardians and poor, in order to guarantee the smooth running of things. The wives of the guardians were also given a task: they were to judge whether the poor were dressed well enough. There also had to be work for everyone: the men could be put to work for the Franeker 'barge makers', the women and children for a wool comber.[32] Franeker did not start to execute Eisinga's plans immediately, but in 1785 a poorhouse was built which was clearly indebted to Eisinga's ideas.[33]

Meanwhile Eisinga's wool-combing business began to blossom. He wrote reports on this in 1811/1812, which serve as a good guide.[34] First, he assessed the most prosperous times of his business had been around the 1770s and 1780s. During that time he made the best turnover, processed the largest amount of wool, and had the highest number of people working for him. Among his employees Eisinga distinguished four different groups. And although the number of people he employed varied throughout the year (in summer there were fewer, because there was more important seasonal work to be done), there were surprisingly many who found work with Eisinga. In modest times he had (1) three aids and two boys; (2) fifty to sixty 'heads of families' (often women) who did spinning for him; and (3) another fifty to sixty girls doing this same work. Finally, he outsourced work to (4) the city poorhouse, which processed more than half his wool. Consequently, the total number of people working for him would certainly

32 Royal Eise Eisinga Planetarium, Eise Eisinga Archive, no. 9.

33 See Telting, *Register*, no. 2061.

34 Dragt, 'Over wol'. In his letters, Eisinga makes the necessary additions to this, see Eisinga to Scheltema, 20 July 1818, UVA, cf. Scheltema, 'Eise Eisinga', p. 195.

have been more than 200 per year, a number that could rise to around 400 in very productive years.[35] Paradoxically, this large business offered Eisinga the chance to spend time on things like his political and social endeavours, while also giving him the opportunity to immerse himself in astronomy.

The wool comber Eisinga emerges as a socially committed and politically active entrepreneur. By the mid-1770s he had not only the training, but also the funding and the social position to spend a lot of time on a new project. It was these circumstances that created the possibility for Eisinga to excel as a devotee of science, as would become clear when he started working on his larger astronomical device.

The Planetarium

Eisinga's planetarium was conceived and built between 1774 and 1781.[36] It resulted in a fully working model of the solar system in his living room. The main model of the solar system is built on a scale of 1:1,000,000,000, which means that 1 cm represents 1 million kilometres. Mercury, Venus, the Earth, Mars, Jupiter, and Saturn (Uranus was to be discovered in the year Eisinga finished his planetarium), orbit the Sun and the Moon orbits the Earth, all in real time. The planetarium gives an abundance of extra information on the position of the stars, the Moon, and all kinds of celestial events such as lunar and solar eclipses. It is operated by a single pendulum clock. As a complete ceiling planetarium, it is considered to be the oldest still in continued operation (Illustration 4).

One of the intriguing questions is why Eisinga decided to build this planetarium. The traditional explanation is that a conjunction of five celestial bodies gave the pretext for him to do so. This story has been told and polished so often that by the early twenty-first century Eisinga is sometimes portrayed as a Galileo-like figure who takes on superstition as well as theology. This is an exaggeration and therefore a closer look at the origin of this tale is warranted.[37] This will give us insight into both the

35 Dijkstra, *Hemelbouwer*, pp. 26–31; see also Dijkstra, 'Eise Eisinga', pp. 6–15.

36 Wopke Eekhoff, who knew both Van Swinden and Eisinga personally, mentioned the years 1773–1780 in the subtitle of a reissue of Van Swinden's *Beschrijving* (1851). Neither in Eekhoff's correspondence, nor in other sources, have I been able to find out why he does this. Unless other sources turn up, it seems to me that this must have been a mistake.

37 See Dijkstra, *Hemelbouwer*, p. 12, where I lean heavily on Breuker, 'Acht maaie', pp. 26–44 and Breuker, 'Eelko Alta', pp. 64–66.

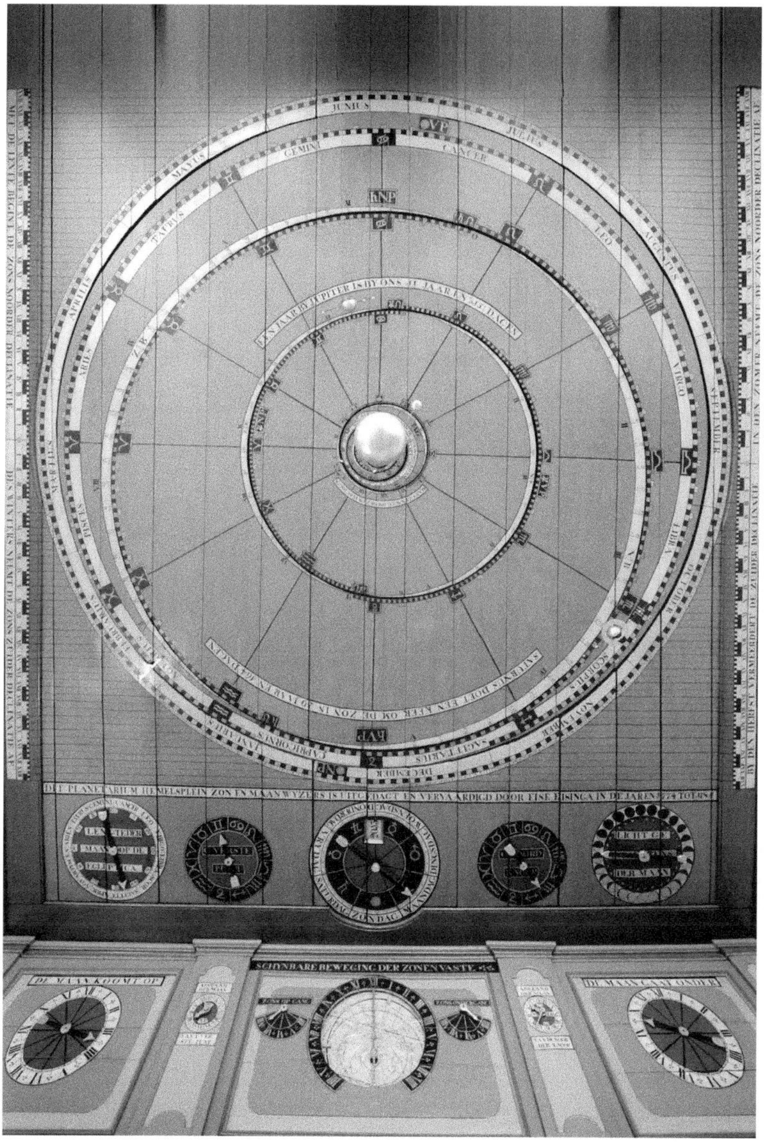

Illustration 4. A composed view of the ceiling and the most important dials of Eisinga's Planetarium, 2012 (*Collection Royal Eise Eisinga Planetarium Franeker*)

reasons for building the planetarium and Eisinga's background working as a devotee of astronomy.

The oldest source on the reason why Eisinga built his masterpiece is a letter written by Van Swinden dated 17 March 1780. Van Swinden would rework his story and even base a well read and often used publication on the planetarium on this first letter. The letter was sent to Jean-André Deluc, only weeks after Van Swinden had first visited the planetarium. In this letter, he writes:[38]

> A prophet is the reason we have this beautiful machine. He predicted that the world would end on 8 May 1774, because multiple planets would be in conjunction, and he wrote a booklet which alarmed everybody. Our citizen calculates by using La Hire's tables this conjunction and concludes that the world will persevere: because it will exist, he says to himself, I will go and make a machine which can show me at any moment what happens in the sky. I will make it in my free time, and it will take me seven years to finish.[39]

These remarks are somewhat tongue in cheek, with the prophet little more than a fun story – the planetarium is the real subject. The humorous tone seems to be enhanced by the biblical reference of taking seven years to build the mechanical universe. As we will see, Van Swinden's correspondence on the planetarium would prove to be very influential. His letter seems to have been designed to spread the news of the curious machine, while granting him the attention as the discoverer of this remarkable invention. So, there were many reasons to tell this tale, but there were other reasons to build the planetarium.

Unfortunately, primary sources are scarce and cryptic. One of the few exists as remarks by Joast Hiddes Halbertsma (1789–1869), a nineteenth-century Frisian scholar, historian, and man of letters. His comments show that the planetarium was a joint project between Eisinga and his father and reveal a whole different motive that the conjunction of 1774. Halbertsma wrote about Eisinga mainly on the basis of hearsay, although he also met him personally on occasion. Halbertsma claims to have heard some striking details about the origin of the planetarium from Idsert Aebinga van Humalda (1754–1834), a long-time governor of Friesland who grew up in the same village as Eisinga. Humalda pointed out that Eisinga 'had been preceded by his father in the

38 Van Swinden, *Beschryving* (1780), pp. 8–9; cf. Terpstra, *Friesche sterrekonst*, p. 228; see also Dijkstra, *Hemelbouwer*, pp. 62–65, 88–91.

39 Jean Henri van Swinden to Jean-André Deluc, 17 March 1780, ULY, Correspondance Deluc: 'Un visionaire est l'occasion de cette belle machine. Il avait predit que le monde périrait le 8 de mai 1774 parce que plusieur planetes seraient alors en conjonction, et il ecrit un livre qui allarme tout le monde. Notre bourgeois se met à calculer cette conjonction par les tables de La Hire et conclut que le monde restera: puis qu'il reste dit-il, je veux faire une machine, qui puisse me'indiquer à tout moment ce qui se passe dans le ciel: Je la ferai a mes heures de loisir, et il me faudra sept ans pour l'achever'.

mechanical hobbies', a claim which is recorded in other sources. He also claimed that it was the father who 'gave his son the idea for a planetarium'.[40]

These remarks gain credibility when we consider that Eisinga, too, mentions his father as an important source of inspiration. In 1818 he writes that Jelte Eises was also much involved in astronomy and even in manufacturing instruments that would clarify the working of the heavens. For example, the father had a homemade rotating instrument in his garden through which the position of the Sun, the Moon and the stars could be determined. He had constructed this instrument because he had noticed an interest in Eise for astronomy.[41] There was a clear family connection in this way of (pursuing and) manufacturing elaborate astronomical instruments. In addition, we know that Jelte Eises was involved in all kinds of other technical matters. He built a sailing dinghy and partook in repairs of a large church organ. Jelte clearly had time to spare, just like his son would later have.[42] The question why Eisinga started building his planetarium should not be answered by looking what was written in the newspaper, but should be sought in the possibility that presented itself. The collaboration with his father gave him the chance to start this project.

For instance, Eisinga reports that between 1768 and 1774 he had no time to occupy himself with astronomy, because he was setting up shop.[43] Apparently this changed in 1774.[44] The successful foundation of his business gave him the opportunity to begin work on a new project. Eisinga was not left to pursue it by himself (which has often been inferred in more popular histories of the planetarium). His brothers and his father helped him build it. For example, Eisinga writes about his father, who at that time had started building a cabinet organ:

> In this my father was rather hindered by my idea of making a planetarium, as we soon realised that this would require a very special lathe in order to turn the necessary axles and discs. To this end my father manufactured

40 Note by Halbertsma in the author's copy(!) of Van Swinden, *Beschryving* (1780), Tresoar, sign. 520 Wk. It has to be noted that Halbertsma's claims were recorded at a very late stage. It was not until 1826 that Halbertsma heard about them, and it was not until 1853 that Halbertsma published about them, which was almost eighty years after Eisinga started building his planetarium.

41 Eisinga to Scheltema, 20 July 1818, UVA.

42 It is clear that father Jelte had lots of free time in which he could do this work. Because Scheltema seems to have interpreted Eisinga's letters rather loosely, an image has even arisen that father Jelte was professionally engaged in the building of instruments and the laying out of gardens. See Scheltema, 'Eise Eisinga', pp. 189–90; cf. Dijkstra, 'Het galjootscheepje uit Dronrijp' and Eisinga to Scheltema, 20 July 1818, UVA.

43 Eisinga to Scheltema, 20 July 1818, UVA.

44 Eisinga to Scheltema, 6 September 1818, UVA. Cf. Van Swinden, *Beschryving* (1780), p. 9.

> the required lathe and on this instrument on my *planisphere*, measuring 28 inches diameter, turned the planet axles and some other discs according to my models.[45]

In the spring of 1775, Eisinga borrowed 500 guilders, a large sum of money, from a Franeker moneylender. It seems likely that he used this money to pay for the construction of the planetarium.[46] A year later, in 1776, the planetarium was ready for Eisinga to paint the first so-called 'Yearplank'.[47] From this we can conclude that within two years the planetarium was taking the shape we know today. At some point during this period, he looked for a clockmaker, who manufactured some crucial parts for him.[48] In 1778, when Eisinga's political career really took off, the planetarium was ready to run.

The spring of 1778, Eisinga published *De Vriesche koopman* (The Frisian merchant), a booklet with tables of different local taxes, very useful for a travelling salesman. In the preface, Eisinga dated it with a long description of the position of various planets at a certain moment, to which he added:

> The reason why I placed this singular date hereunder, was because my present half-finished mobile Astronomis Planetsphaerium, which I am working on, showed and pointed out to me these positions of the planets as if with a glance of the eye, and to further exhort and arouse You [the reader], who finds some pleasure and amusement in the astronomical movements, to investigate the ways of that heavenly ensemble, in order to obtain closer and more extensive knowledge of the great, glorious and wonderfully created universe.[49]

45 Eisinga to Scheltema, 6 September 1818, UVA.

46 In the years 1770, 1772, and 1775 Eisinga three times borrowed a large sum of money, the first time £1000-, the second time £999- (in order to repay the first loan), and the third time £500. Tresoar, Nedergerecht Franeker (12-12), inv. no. 311.

47 Van Swinden, *Beschrijving* (1824), p. 143. See also Eisinga's 'Beschrijving', p. 253.

48 Van Swinden, *Beschryving* (1780), p. 9. See also: Engelman to Van Swinden, 5 October 1780, UBL, PBL 755.

49 'De reeden waarom ik deese singuliere datum hier onder geplaatst heb: was om dat mijn tans onder handen zijnde half gemaakt beweeglijk astronomis planeetsphaerium, mij deese stand plaatzen der planeeten als met een opslag van 't oog vertoonde en aanwees, en om UE. die eenige lust en vermaak in de astronomische beweegingen hebt, verders aan te spooren en op te wekken, om de wegen van dat heemels geselschap na te vorschen, ten einde nader en uitgebreider kennis te bekomen, van het groote heerlijke en wonderbaare geschapen heel-al'. Eisinga, *De Vriesche koopman*, preface. There are (possibly several) later editions of this booklet that have appeared where this preface is missing. The date is 6 June 1778, in the afternoon.

This is the first time of which we are certain that Eisinga spoke about his planetarium in public. And from the start he seems to have the ambition to make his work known to the world – it was an explicit invitation to come and see the planetarium. This helps us explain why the planetarium outperforms a textbook, a drawing, or other depiction of the solar system: it can provide insight into the movements of the celestial bodies.

The booklet with the tables is special not only on account of the preface. It will certainly have served as a useful aid to travelling salesmen and it marks the year 1778 as an exceptionally busy one in Eisinga's life. It was possibly one of the best years for his business, he was elected guardian of the poorhouse, he drew up the plan for its refurbishment, and published *De Vriesche koopman*, while work on his planetarium was also progressing. It seems that for Eisinga, the sky was the limit.

According to some authors, Eisinga worked on his planetarium virtually in secret. Even during Eisinga's lifetime, Jan Brouwer worded this as follows (1824): 'How he hid and wrapped himself, as it were, in his own quiet virtue and homely pleasures! Without Van Swinden [...] it would have remained in obscure oblivion, or would soon have sunk'.[50] How this relates to Eisinga's own announcement in *De Vriesche koopman* is left open by Brouwer.

This style of description, through which a man's deeds are praised as grand, but his personality as modest, fits well with the times: it would soon be seen as a romantic ideal. In addition, it seems to have suited the person of Eisinga, whose gentle and calm nature is often remarked upon by visitors as very striking. Be that as it may, Eisinga was a citizen with ambition and presence, who managed to balance his responsibilities as an entrepreneur, his societal role in the community, and his talents as a devotee to astronomy. The question why he built his planetarium should thus be answered by pointing to Eisinga's skills, the help he received from his father, the time he had on his hands, and the money he could borrow. He had the means and the opportunity to construct the planetarium.

The Discovery of Heaven

On 22 February 1780, Van Swinden visited Eisinga. By that time the latter had been working on his planetarium for almost six years. Meanwhile it was running, the hands of the clock were installed, but it was not yet finished: no colours had been applied yet, everything was still 'marked with chalk'.

50 See Jan Brouwer's preface in Van Swinden, *Beschrijving* (1824), p. v.

The following year, Eisinga was to paint it, together with his brother.[51] Van Swinden went straight home, where he made a description of the machine as best he could. These notes have been preserved.[52] They show that in the weeks after his first visit, he was far from idle. He went back numerous times, made lists of older orreries, checked Eisinga's calculations, and wrote about the planetarium with various correspondents at home and abroad.

Van Swinden was a very well-connected academic with correspondents all across Europe, thus his preserved correspondence offers a wonderful insight into his amazement at the planetarium. Furthermore, tracing his letters and the way they were copied provides a glimpse of how fast news could spread throughout the world. In his *Beschryving* (Description) Van Swinden related which well-known scholars he informed. He wrote to the Russian Dimitri Gallitzin, the Swiss Jean-André Deluc, and the Frenchmen Jean Gaussen and Father Louis Cotte.[53] Cotte forwarded Van Swinden's letter to a French scholarly journal, where it was published in its entirety. Subsequently, the news about the planetarium attracted the attention of readers from all over the world. Within months, the letter not only appeared in other French journals, but was also translated in English, Italian and Dutch journals.[54] In Germany, summaries were published.[55] In 1782 no less than five American revolutionary newspapers published copies of the letter to 'Father Cotte', these were *The Independent Gazetteer, The Providence Gazette and Country Journal, The Independent Ledger and the American, The Salem Gazette*, and *The Massachusetts Spy*. Some published the letter on their front page. Within two years, Van Swinden's initial report had enabled people all over the world to become acquainted with the Franeker Planetarium.[56]

Meanwhile Van Swinden seems not to have mentioned the planetarium to correspondents in the Dutch Republic, until his *Beschryving* was published in the summer of 1780. That publication again drew the attention of national and international periodicals – while the initial letter was

51 Van Swinden, *Beschrijving* (1824), pp. 143–44.

52 KB, Van Swinden Collection.

53 See for instance Van Swinden to Gallitzin, May 1780, UBL, PBL 750; see for more examples Dijkstra, *Hemelbouwer*, pp. 88–91.

54 Examples can be found in *Observations sur la physique* (July 1780), pp. 456–59; *The Monthly Review or Literary Journal* (July–December 1781), pp. 150–56; *Gazetta di Parma* (2 March 1781), pp.70-73.

55 *Gothaische gelehrte Zeitungen auf das Jahr 1782*, p. 80.

56 In 1783 Van Swinden received questions from English journalists about the Planetarium. See Eekhoff, 'Het leven', p. 25.

also still circulating. Various of Van Swinden's correspondents thanked him for this new publication, which he did send to his Dutch friends. The Haarlem physician J. Engelman said that he had loaned it to an 'instrument maker' – who apparently also knew Eisinga personally. This was cause for Engelman and Van Swinden to maintain further correspondence about planetariums, whereas previously they mainly wrote about meteorological issues.[57] Franeker-born Sebald Justinus Brugmans (1732–1789), a renowned botanist and physician, showed himself to be particularly impressed by Eisinga's ingenuity on the basis of Van Swinden's publication.[58]

Van Swinden also sent a copy of his *Beschryving* to the Haarlem academic Martinus van Marum (1750–1837), one of the foremost scholars of the time. With Van Marum, too, Van Swinden usually maintained contact about other matters, such as thermometers, hygrometers, and electrostatic generators. The Haarlem director of the Holland Society of Sciences did not waste any time and visited Van Swinden in the summer of 1781. Of course, they went to the planetarium, accompanied by Van Marum's wife.[59]

Regrettably, it is precisely from the first years that many of the planetarium's visitors' data are missing – the guest books lack several dozen pages. It is therefore difficult to trace who did and who did not travel to Franeker in order to view Eisinga's work with their own eyes. Nevertheless, it is clear that Van Swinden's favourable descriptions attracted people. Eisinga himself estimated that between 1780 and the summer of 1787, the planetarium drew more than a thousand visitors. Especially during the summer of 1785, on the occasion of the celebration of the second centenary of Franeker University, a prominent group arrived.

News about the planetarium had struck like lightning. In one fell swoop it turned Eisinga into a famous citizen. Van Swinden saw wonderful opportunities for the University of Franeker, campaigned for a second planetarium, and applied for recognition from the Frisian provincial regents. This recognition came in the form of only a silver coffee and silver tea pot, a visit to the planetarium, and a small sum of money. This minimal recognition was an insult to Eisinga, whose story appealed to people throughout Europe and

57 Engelman to Van Swinden, 5 October 1780, UBL, PBL 755; Van Swinden to Engelman, 21 November 1780, UBL, PBL 755.

58 Brugmans to Van Swinden, 24 November 1780, UBL, PBL 755; see also Van Heiningen, *The Correspondence*, p. 28.

59 Van Marum's visit is missing from Eisinga's guest books, but it was at one time included: see Eisinga to Scheltema, 6 September 1818, UVA; the fact that Van Marum visited Van Swinden can be read in Van Swinden to Bruining, 30 August 1780, UBL, PBL 755.

North America. It was clear that people wanted to learn about bourgeois who were educated by their own force of will, elevating themselves independently of pedigree, lineage, or noble descent.

Chased and Imprisoned

Eisinga continued as entrepreneur and politician, and maintained his devotion to mathematics and astronomy even after he had finished his planetarium. In 1783 he found time to compile a large manual for his planetarium. And in the following years he became ever more politically active. This led to his involvement with a people's revolt in 1787 in his town of Franeker. The Dutch 'Patriots', who found inspiration in the American Revolution, and later the French Revolution, tried to rise to power in the late summer of 1787. In Friesland, too, they took up arms, and chose Franeker as their new centre of government and Eisinga as one of their leaders.[60]

Although the Frisian patriots managed to seize power, they were unable to hold on to it. By the end of September, they had to leave the city under pressure of the king of Prussia, who came to the aid of the political Dutch establishment. Some of the leading revolutionaries fled to France, where they found shelter in Saint-Omer. Eisinga fled to Germany, where at first, he stayed in the small towns of Steinfurt and Gronau. In Steinfurt, he shared a small room with his townsman Jacobus Scheltema (1767–1835). Other Franeker revolutionaries found shelter there, too.[61]

Together, Eisinga and Scheltema made a paper model of the planetarium, which they showed to the count of Steinfurt. Eisinga had even brought his guest book with him during his exile, and the German count proceeded to sign it as if he had visited the planetarium.[62] Eisinga also paid regular visits to Lambertus Nieuwenhuis (1741–1810), another devotee of astronomy. Nieuwenhuis lived in Enschede and had visited the planetarium in or around 1783.[63] Eisinga left a lasting impression on him and Nieuwenhuis in turn found a follower in Coenraad ter Kuile (1781–1852), who was to build Tellurium, a mechanical model of the Sun, the Earth, and the Moon.[64] This functioned somewhat similarly to the Franeker Planetarium: it did not

60 I describe these events in more detail in Dijkstra, *Hemelbouwer*, pp. 111–17.

61 Botke, *Boer en heer*, 568; Lindeboom, 'Een patriots predikant'.

62 Eisinga to Scheltema, 6 September 1818. UVA.

63 Nieuwenhuis to Van Swinden, 9 April 1803, UBL, PBL 755; see also Havinga et al., *Planetariumboek*, pp. 199–200.

64 Havinga et al., *Planetariumboek*, pp. 248, 368–69.

take long for travellers to Enschede to pay a visit to this Tellurium.[65] Even in exile, Eisinga's planetarium brought him social standing.

Eisinga was both worried and bored during his time in the German lands. After a year, in 1788, his wife died. Their children, who were staying in Franeker, were looked after by their family, as were Eisinga's belongings. This meant that his family also took care of renting out the planetarium – as a residential building. In the few of his letters that have survived, an image emerges of a homesick man, who filled his time with mathematical and astronomical riddles. He pleaded his innocence to his family and plans were made to plead for amnesty, but that never materialised.[66] There was even talk that Eisinga could be patronised by the French court, but that too never came about.[67]

At a certain point Eisinga decided to travel back to the Netherlands. He felt this was safe because the Frisian government had no jurisdiction outside its own province. He thus considered himself safe in the province of Groningen. Here he lived just across the Frisian border in Visvliet, a small village.[68] In July 1791 he bought a little house, and started a new, small, wool combery. In a letter to Van Swinden, he noted that hardly anyone doubted he was safe, 'although up till now I have not been able to acquire absolute certainty'.[69]

He was wrong. In May of the following year, Eisinga was arrested and brought to Leeuwarden, in Friesland, to be tried. A lengthy process began, the outcome of which was far from certain. Initially, the prosecutor even seems to have alluded to demanding the death penalty, because not only was Eisinga a member of the Franeker Defence System, he also actively contributed to the revolution. He ran the administration, for example, and he was said to have ordered troops, helped build defences, and signed an objectionable public letter. Eisinga confessed his membership and the fact that he at times had lent a helping hand, but he denied the other points. The judges soon seemed to accept this, to Eisinga's relief. He now expected that he would be detained for the remainder of the period of the trial, and that following the proceedings he would be subject to 'banishment outside the province for some time', and thus the matter

65 Halbertsma, for example, paid a visit. Friendly communication from Alpita de Jong.

66 Dijkstra, *Hemelbouwer*, pp. 121–23.

67 Eisinga to Court Lambert van Beyma, Gronau, 29 May 1790, Tresoar.

68 Eisinga's letters can be consulted at Tresoar. For the edition, see: Havinga et al., *Planetariumboek*, pp. 169–96. See also: Botke, *Boer en heer*, p. 118.

69 'egter heb ik tot nog toe geen volkomen Seekerheid kunnen krijgen', Eisinga to Van Swinden, 30 July 1790, UBL, PBL 755.

would be resolved. And so it happened. Eisinga was convicted – and this seems rightfully so – but the more serious charges were all declared inadmissible.[70]

On 27 April 1792, Eisinga completed his time in prison and was granted two days to leave Friesland. He immediately left for Dronrijp. The next day he visited his beloved planetarium one last time, before leaving for Visvliet, where he picked up his wool-combing business again, remarried, and expected to serve the time of his banishment.[71] His time in Visvliet passed without too many significant events, although even here he welcomed visitors from time to time who wanted to meet the famous builder of the Franeker Planetarium.[72]

Within fifteen years the planetarium had brought him fame and recognition. This had not always been a blessing – as he was so well known, his prosecution was one of the harshest of all the revolutionaries of 1787. This may have been a sign that the claim to power and recognition made by Eisinga and his bourgeois patriots is also expressed through the planetarium. It makes Eisinga's celestial clock a form of self-representation. Be that as it may, it would take time before their movement had gained enough momentum to succeed in actively overthrowing the sitting government.

Recognition for Time Devoted to Astronomy

With the Batavian Revolution of 1795 the Patriots managed to gain power in the Netherlands after all. They succeeded with the help of the French revolutionaries. This opened the road back to Franeker for Eisinga, and his banishment was cut short. In the following years, Eisinga refurbished the astronomical timepiece, put everything back into operation, and from 1797 he once again received visitors. Where Eisinga's fame had been a risk during his years in exile and prison, he now reaped the rewards again and put this social capital to good use.

Immediately after his return to Friesland, Eisinga also took on prominent political tasks. He did not become a political leader who stirred up crowds, but fulfilled the role he was supposed to have played in 1787: he took care of the administration and financial matters. As a representative of Franeker, Eisinga took a seat on the provincial board and was briefly even a delegate

70 Tresoar, Archief Hof van Friesland, Eijse Jeltes Eijsinga, inv. no. 4657.

71 Eisinga to Van Swinden, 30 April 1792, UBL, PBL 755.

72 Martens, *Ik heb het groote doel*, pp. 62–63.

to the first national parliament of the Netherlands.[73] The political situation between 1795 and 1800 can only be described as highly turbulent, with constant power changes and shifts in allegiances and factions.[74] Eisinga had a remarkably long run as a politician, made few enemies, and continued to perform the administrative tasks he was assigned.[75]

In 1797, the new-found administrative position resulted in one of the most important assignments of his life. Together with Yme Freerks Tigchelaar from the village of Makkum, Eisinga drew up a plan for the future of the University of Franeker. He did this in the context of his administrative responsibilities, as a member of the Provincial Board and perhaps particularly as a member of the finance committee. The pair were to ensure that, in future, this particular university gained greater social significance. The plan presented by Tigchelaar and Eisinga denoted a remarkable mixture of an academy for common people, with room for both enlightened and more romantic ideals. Education should largely take place in Dutch and no longer in Latin. Theology, until then the crown jewel of the academy, was moved to the margin. Subjects such as history and geography gained importance, as did, obviously, mathematics and astronomy. On the subject of mathematics, Tigchelaar and Eisinga wrote: 'Mathematics is a science of necessity, utility and influence in general, and of which, if taught well, the pleasantness of practice for the young also will not ever have any limits whatsoever'.[76]

They even spoke of mathematics as a woman with whom 'thou shalt be in love' and who shall be 'thy tenderest and dearest friend'.[77] Their ode to mathematics was followed by an ode to mechanics. The authors elaborated this on the basis of 'mill construction'. If anywhere the usefulness of this mathematics-related subject became clear, it was in the construction of mills. Perhaps here the experience of building a planetarium resonated: after all, it had many similarities with a very small, but exact mill.

Finally, astronomy received the highest praise. When giving the assignment to Tigchelaar and Eisinga, the University Senate made an ostentatious attempt to temper the importance of astronomy somewhat in a letter in

73 Brood, *Homines novi*, entry: Eisinga.

74 See, for example, what happened to the Leeuwarden representative Wopke Wopkens, see: Dijkstra, 'Het stokske'.

75 See also Kuiper, *Een revolutie*, p. 182.

76 'De Wiskunst is eene Weetenschap welker noodzaakelijkheid, nuttigheid en invloed over het algemeen, en welker aangenaamheid van beoeffening voor de Jeugd tevens, zo zij wel wordt Onderweezen, in dat alles volstrekt geene paalen kent nog ooit zal hebben'. Tigchelaar and Eisinga, *Plan*, p. 23.

77 'uwe tederste en waardste hartsvriendin'. Tigchelaar and Eisinga, *Plan*, p. 23.

which it complained about the state of affairs. Students could 'be encouraged to also attend the lessons in Nature and Astronomy, provided that this does not rob them of too much time', according to the university administrators.[78] Tigchelaar and Eisinga did not take much notice of this and sang the praises of astronomy in their plan. Not only did the practitioner of astronomy learn to understand the world, it was also a means of discovering the greatness of God.[79] This reflected the romantic ideals, as much those of the physico-theologians of the seventeenth and eighteenth centuries: through nature, the existence of God was learned. It was a line of reasoning that could also be found in the planetarium, where on a bar above the workplace it said, 'Remember that God is always in your presence'.[80]

The plan drawn up by Tigchelaar and Eisinga fitted in nicely with the general reshuffling of the university's position in society. All over Europe such issues were thought, written, and quarrelled about. In the Netherlands, one of the opinion leaders was inevitably Van Swinden. He had been installing his students in crucial positions since the 1780s. And although there is no known correspondence from him about this specific plan, his ideas, too, were reflected in it. For example, emphasis on the social usefulness of the academy was one of the key points of Van Swinden's programme, which was, as Klaas van Berkel showed elsewhere in this volume, a larger European movement. Tigchelaar and Eisinga were consequently working on a very modern project.

Contemporaries were not very much impressed. Some years later, a scathing criticism appeared in *Vaderlandsche letteroefeningen*.[81] One commentator called the plan 'incomprehensible semi-inspired drifting', and recognisable 'as a sketch of an encyclopaedia by an uneducated person'.[82] However, that was a criticism aimed at Tigchelaar, whereas Eisinga was praised by the same critic for 'the excellent art and ingenuity, and diligence' with which he made his planetarium. More praise was conferred by others, also on Tigchelaar. The Franeker professor Everwinus Wassenbergh (1742–1826) was

78 Tigchelaar and Eisinga, *Plan*, p. 13.

79 Tigchelaar and Eisinga, *Plan*, pp. 27, 35–39. Incidentally, other things are also reflected in this: the authors claim, in accordance with Christiaan Huygens, that elsewhere in the universe other beings must also exist that praise and honour God. Tigchelaar and Eisinga, *Plan*, p. 39.

80 Until recently this board hung next to the automaton which drives the Planetarium, where it was placed long after Eisinga's passed away. Cf. Langereis, *Machineman*, 146; Langereis clearly is not familiar with this. The line itself is taken from the main filing cabinet in Franeker town hall from the seventeenth century.

81 Van der Kroe, *Vaderlandsche Letteroefeningen*, vol. 2 (1803), pp. 592–95.

82 See: Letter from Hendrik Willem Tydeman to Johan Meerman, 2 March 1807, in Breuker, 'Een overzicht', p. 145.

very outspoken in this respect, and later historians too regard Tigchelaar and Eisinga's plan as the last time that the University of Franeker had the wind in its sails, thanks to the efforts of the authors.

From 1805 onwards Eisinga's role in national and regional politics ended. He retired to Franeker, where he kept welcoming visitors to his planetarium. However, he did not relinquish all his political influence: he kept using his experience and fame for his townsmen. Data, letters, and reports from the provincial archives of the province of Friesland as part of the Kingdom of the Netherlands (from 1813) reinforce this picture. A number of times, Eisinga had contacts and collisions with provincial administrators. From this it becomes clear that he stood up for the interests of his fellow townsmen and, if necessary, even battled authority on their behalf.[83] Again and again Eisinga comes across as a straightforward, politically active man, who was not afraid to use the position he had gained through his fame as builder of the planetarium.

In 1816 Eisinga was recognised by the Dutch government for his work, but presumably especially for making his planetarium and keeping it open for the public. This recognition also saw him nominated for the award of 'Brother in the Order of the Dutch Lion'. The award consisted of a medal, and a yearly allowance of 200 guilders. On top of this, King William I paid a visit to the planetarium in 1818. Eisinga was extremely proud of this. He wrote:

> I also had the flattering pleasure and the great honour to receive in my humble home on 30 July last our beloved King with his son Prince Frederick. His Majesty viewed everything carefully. Even the inner parts of the mechanism, crawling into the narrow and confined space in order to contemplate the cogs and the uneven movements of the moon. His Majesty tarried close to three quarters of an hour in our house, and seemed to be well pleased.[84]

The king's visit and Eisinga's contacts with his old fellow villager and governor of the province Idsert Aebinga van Humalda turned out to be of

83 Archief Provinciaal Bestuur van Friesland 1813–1922, in particular file 6102. This archive contains more than thirty files mentioning Eisinga and members of his family.

84 'Ook heb ik het streelend genoegen en de groten eer gehad onsen beminden koning met [...] prins Frederik [...] in mijn geringe woning te mogen ontvangen. Zijn Majesteit heeft alles nauwkeurig bezigtigd. Zelvs de binnendeelen van 't raderwerk, kruijpende in de enge en bekrompen ruimte om de raderen en de onegale [=onregelmatige] beweegingen van den maan te beschouwen. Zijn maijesteit heeft nagenoeg ¾ uur in ons huis vertoefd en scheen zeer wel voldaan te zijn'. Eisinga to Scheltema, 20 July 1818, UVA; cf. Scheltema, *Eise Eisinga*, p. 234.

crucial importance for the purchase of the planetarium by the state, which secured the fate of the planetarium and the future of Eisinga's family.[85] That his descendants never managed to attain the influential position that Eisinga himself had had is a bit wry. It seems that care for the planetarium curbed further social progress for the family.

It is clear that around 1820 Eisinga was a man of stature in and outside Frisian society. He had also made a name for himself as an astronomer. The Groningen almanac writer Abel Jans Sleuver (1786–1847) (also skipper, shopkeeper, and livestock farmer) boasted of his correspondence with Eisinga.[86] When in 1820 an exceptional partial solar eclipse took place, Eisinga went to discuss this with the astronomer Arjen Roelofs (1754–1828), together with 'a few professors' and Governor Humalda.[87] It was rather reminiscent of the visit to Wytze Foppes at the time of the transit of Venus in 1761.

Although by now Eisinga was approaching his eightieth birthday, he still made time to practise astronomy and mathematics. Even as an old man, he still reaped the rewards of devoting so much time to these sciences.

A New Planetarium

Over time, many people became enthusiastic about the planetarium, and a small group tried to get a second construction off the ground. It is inevitable that the first attempt was made by Van Swinden: in 1783, he proposed to the Senate of the University of Franeker to let Eisinga build a planetarium and, following that, immediately also proposed to charge Eisinga with the care of it. In addition, Van Swinden himself was to publish a description of this second planetarium, this time an illustrated description. The Franeker University Senate did not approve of the idea.

A much more detailed plan came from Eisinga himself. In 1790, when he was in Gronau, he wrote to his family and friends that 'in sleepless nights an excellently beautiful Hemisphaerium & Planetarium has come to my mind, much simpler, much more perfect, and with nature represented much more truly than in ours'.[88] He elaborated on this in detail, and asked his

85 This episode is well researched by Mulder de Ridder, 'Eise Eisinga'.

86 Cf. Botke, *Boer en heer*, p. 310. He visited the Planetarium twice, on 14 January 1821 and 23 July 1826.

87 Van Albada, *Uit de oude en nieuwe doos*, pp. 116–17.

88 'in Slaapelose Nagten in de gedagten is gevallen een Uitmuntend Schoon Heemelsphaerium & Planetarium veel Eenvoudiger veel volmaakter en met de Natuur die 't Selve verbeeld veel overeenkomstiger als 't onse is', quoted in Havinga, *Planetariumboek*, p. 170.

brother to offer this new planetarium to the States of Friesland, by way of compensation for his role in the revolt of 1787. It was to no avail. The aversion towards Eisinga was too fierce, and the Frisian regents did not reply to his offer.

However, this did not mean that the plan for a new planetarium was abandoned. Eisinga pursued it, and in 1790 from Visvliet he reported to Van Swinden that he had made good progress with the plans, although he met with difficulties in inserting Uranus; the distance from Uranus to the other planets was so great that he expected to need a second timepiece for it. He was also nearing the limits of his mastery of mathematics: the continued fractions with which he calculated the proportions of the existing planetarium did not suffice for his new plans: 'I think that such a problem can be solved by means of algebra, which I do not master sufficiently'.[89]

The challenges posed by a new, finer planetarium were becoming more urgent when in 1791 Eisinga was imprisoned in Leeuwarden. He was visited there by Salomon de Monchy (1714–1794), a physician from Rotterdam who had been one of the founders of the Batavian Society. He was willing to use his influence to establish a planetarium in Rotterdam. Earlier that year, De Monchy had checked with Van Swinden the possibility of having a new planetarium built in an existing building.[90] Based on Van Swinden's suggestions he visited the blockhouse in Leeuwarden to exchange ideas about this with Eisinga himself. The latter was immediately enthusiastic. He described his plans and even drew up a budget. Eisinga thought that a new planetarium would cost some 6,000 guilders, including materials and even including a director's salary for a period of six months.

Eisinga envisioned a dome in which the planetarium would be placed. By means of a staircase, visitors could get some kind of overwhelming impression of the universe. It is possible that the plans presented by Eisinga were somewhat more extensive than De Monchy had in mind. The man from Rotterdam had clear limitations, for example, while Eisinga wrote a proposal for which 'money or costs were no object'. De Monchy died in 1794, but by that time the new planetarium had not been discussed for a while.

In 1800, the plans were dug out again when the Amsterdam group Felix Meritis showed an interest. Again, Van Swinden was involved and again there seemed to be a fair chance that the second planetarium was to be realised. This time it was professor Gerrit Hesselink who advocated the creation of a planetarium such as Eisinga's. One of the members of Felix Meritis,

89 Eisinga to Van Swinden, 27 August 1791, UBL, PBL 750.

90 Eisinga to Van Swinden, 27 August 1791, UBL, PBL 750.

Jacobus d'Amour, immediately laid down 1,000 guilders to realise this plan. But because the plans differed profoundly from what had been imagined, construction was abandoned. Eisinga's plans were shelved once again.[91]

In 1808, the plans were dusted off when King Louis Napoleon was talked into agreeing to a new planetarium, to be built by Eisinga. But again, the idea did not materialise.[92] The next person of whom it can be determined with certainty that he seriously enquired after a second planetarium was Jacobus Scheltema in 1818. By that time, Eisinga's plans turned out to have been developed further again. He claimed 'now to be in a position to make a similar work, that would surpass my first in many respects, but my years are at present so advanced that the end may soon be there'. When Scheltema insisted, it became clear that Eisinga would really like to see two planetariums in one building. This would provide him with the opportunity to accommodate newly discovered celestial bodies in the new planetarium:

> Because of the great distance of Uranus, the planetary system should be divided in two, so that on the first one Mercury, Venus, the Earth, Mars, Vesta and Juno could be represented and on the second our Earth's Pallas, Ceres, Jupiter, Saturn and Uranus, the first surrounded by a month and date pointer, the second by a sign and degree pointer for the Sun, beside the first the open setting of the Sun and beside the second the Sun's declinations. If one were to have a location three times as long as it was wide, one could make a Planisphere in the middle between both planetary systems.[93]

As good as the plan sounded, Scheltema too failed to get it approved. In 1824, when Van Swinden's *Beschrijving* was republished, Jan Brouwer (1760–1838) made one last attempt, but by then the belief in a second planetarium had already dissipated.[94] What has remained is the existing work, which meanwhile had a robust story to accompany it. Part of that story was the second specimen that was never built. Eisinga had become a romantic hero.

91 Zuidervaart and Van Gent, *Between Rhetoric*, pp. 58–61.

92 Eekhoff, 'Het leven', p. 32.

93 'Door de grote afstand van Uranus behoorde het planeetgestel in twee-en gedeeld te worden, dan kon men op 't eene brengen Mercuur Venus de Aarde Mars Vesta en Juno en op het 2de ons Aarden Pallas Ceris Jupiter Saturnus en Uranus het eerste omringd van een maand en datum wijser, het 2de van een teeken en graadwijser voor de Zon, neevens het eerste de open ondergang van de Zon en neeven het 2de de Zons declinatien. Als men een locaal had dat 3 maal zo lang als breed was, konde men in 't midden tusschen de beijde planeeten stelsels een Hemelsplein maken'. Eisinga to Scheltema, 17 August 1818, UVA.

94 See Brouwer's introduction to Van Swinden et al., *Beschrijving* (1824).

What Eisinga did manage to do in the meantime was to build a reputation as an instrument maker. He has been known to have made sundials, a table planetarium, and moving star charts. He gave these instruments to good friends, or he sold them to acquaintances.[95] It was not big business, but it served to underline his position as a famous astronomer.

What all these propagandists of the bigger and better planetarium hoped to achieve was fame, by building an educational model of the solar system. Their motives seem to overlap with those who build planetariums today. They aim to teach people about the universe by making visible how the celestial bodies move, while simultaneously hoping to attract the attention of many. Clearly the institutional structures were not yet strong enough to support such developments. In that regard, Eisinga's biggest achievement may have been simply to succeed the first time. He did so without the financial help of any government, church, or patron. For an even larger project for which did need such support, he never succeeded.

In 1828 Eise Eisinga died. By that time, the future of his planetarium had been settled. To preserve it for the general public and for future generations it had been purchased by the Dutch state and was managed and used by the Athenaeum – the successor to the University of Franeker.[96] This was fortunate for the Franeker Planetarium and one of the major reasons why it has weathered the times so well. The governor of Friesland even commissioned a portrait of Eisinga by Willem Bartel van der Kooi, one of the society painters of the period. At the end of this life, he received all the recognition he had craved for so long.

Conclusions

Eise Eisinga was a typical product of his time: Friesland in the second half of the eighteenth century. This in part makes his planetarium a product of the Enlightenment. As many visitors have attested it is undeniably a very powerful didactic model to explain the workings of our solar system. Understanding nature, representations of planets, Sun and Moon, explanation of the course of things, these all fit into this model beautifully. In addition, it is one of the finest examples of what an individual could be capable in the eighteenth century, provided that conditions were favourable. In Eisinga's case, these were the presence of a strong mathematical and astronomical

95 Eisinga to Scheltema, 17 August 1818, UVA.

96 Vreeling, 'Eise Eisinga was hier', pp. 22–23.

culture, the financial position to be able to support such a project, sufficient time to execute it, the determination to carry it through, and most of all the enormous talent required to make the calculations. At the same time this culture also provided an audience to appreciate his talents and his planetarium.

Due to these elements, Eisinga acquired a position in society which would have been hard to achieve in almost any other way. He took full advantage of this position. Sometimes things went wrong, Eisinga himself often felt he was not appreciated enough, but it was his fame and status that gave him trouble during the period of his imprisonment and banishment. More often things went well, as evidenced by his position in Franeker in the nineteenth century, the eventual purchase of the planetarium by the state, and the beautiful portrait painted of him. We have seen that Eisinga had the means to build his planetarium. He had the money to do it, and the skills to calculate everything. Furthermore, he could count on the help of his family, so that during a period when he had the time, he also had the opportunity. All the recognition he received may point to his motive. Eisinga benefited socially from his planetarium – even though he also learned that being famous was not always a pleasure.

The stunning appearance of the planetarium, together with its perfect fit with Enlightenment ideals, turned it into an almost instant success. The planetarium is an expression of useful science, especially since it helped introduce people to astronomy and gave those with knowledge of the subject the possibility to enhance their expertise. The central figure in conveying this message was Van Swinden, one of the first visitors we have a record of, who was immediately overwhelmed by the planetarium. Thanks to his correspondence and publications Eisinga soon became one of the first real romantic heroes of the Netherlands. Over time, some facets of this have been exaggerated and subsequently downplayed again, but they have never completely disappeared. This is partly because Eisinga seems to have the characteristics of the ideal romantic scholar. He moved off the beaten track and did not have the advantage of a classical education. From descriptions he comes across as somewhat eccentric. This makes the tendency to regard him as a romantic devotee very tempting. However, Eisinga was not an autodidact, but received a proper education. Nor was he a simple artisan, but owned a large company with sufficient financial scope. He also managed to climb the social ladder. Furthermore, he started to function in the scientific world, helped to shape the University of Franeker, and was one of the very first Dutchmen to be awarded a royal decoration. Ultimately Eisinga is a much more interesting character than the one who has been cast to fit the romantic mould.

From the nineteenth century onwards, Eise Eisinga's paper trail seems to point in the direction of a brilliant, self-taught, but simple artisan. However, a more elaborate analysis shows that he can be understood as a typical product of his time and his age. Eisinga and his planetarium are not enigmas; they are phenomena that fit within a broader context worthy of greater exploration.

Bibliography

Botke, IJ., *Boer en heer: De Groninger boer, 1760–1960* (Assen 2002).

Breuker, Ph.H., 'Acht maaie 1774: Panyk en ferljochting', *De Vrije Fries* 64 (1984), pp. 26–46.

Breuker, Ph.H., 'Eelko Alta as Patriot', in: W. Bergsma et al., *For uwz lân, wyv en bern: De patriottentijd in Friesland* (Leeuwarden 1987), pp. 59–70.

Breuker, Ph.H., *Eelko Alta (1723–1798). Biografie van een verlicht volksvoorlichter* (Gorredijk 2024).

Breuker, Ph.H., 'Een overzicht van de geleerdheid in Friesland in het jaar 1807', *De Vrije Fries* 91 (2011), pp. 141–48.

Brood, P. et al., ed., *Homines novi: De eerste volksvertegenwoordigers van 1795* (Amsterdam 1993).

Buys, E., *Nieuw en volkomen woordenboek van konsten en weetenschappen*, vols. 1–10 (Amsterdam 1769–1778).

Dijkstra, A., 'Het stokske van Wopkens: Een portret van de Leeuwarder patriot Wopke Wopkens (1730–1807)', *Leeuwarder historische reeks* 8 (2005), pp. 135–51.

Dijkstra, A., *Between Academics and Idiots: A Cultural History of Mathematics in the Dutch Province of Friesland (1600–1700)* [Doctoral thesis] (Enschede 2012).

Dijkstra, A., *De hemelbouwer: Een biografie van Eise Eisinga* (Gorredijk 2021).

Dijkstra, A., 'Eise Eisinga (1744–1828), meester-wolkammer', in: D. van den Beukel et al., ed., *Eise x Pleed* (Leeuwarden 2021), pp. 6–15.

Dijkstra, A., 'The Planetarium and Eise Eisinga', *The Royal Eise Eisinga Planetarium: Nomination by the Kingdom of the Netherlands for Inscription on the UNESCO World Heritage List*, vols. 1–3 (Franeker 2021), vol. 3, pp. 64–95.

Dijkstra, A., 'Het galjootscheepje uit Dronrijp: Eise Eisinga en de lessen die hij leerde van zijn vader Jelte Eises', *Jaarboek 2021. Fries Scheepvaart Museum en Oudheidkamer* ([Sneek] 2022), pp. 5–12.

Dijkstra, A., 'Spectaculaire vondst zet Eisinga in ander daglicht', *Letterhoeke* 19, no. 1 (2023), pp. 8–9.

Dragt, G.I.W., 'Over wol, tegels en Eise Eisinga als "fabrikeur in Vriesche sajet"', in: M. Seffinga, ed., *It doe yn 't hjoed wer woun: Liber amicorum voor Sytse ten Hoeve* (Sneek 2001), pp. 80–90.

Eekhoff, W., 'Het leven van Eise Eisinga en beknopte geschiedenis van zijn Planetarium', in: J.H. van Swinden, *Beschryving van een konst-stuk, verbeeldende een volleedig bewegelyk hemels-gestel, uitgedagt en vervaardigd door Eise Eisinga* (Franeker 1780), pp. 1–43.

Eisinga, E., *De Vriesche koopman, zijnde een uitreekening van alle impostien etc.* (Franeker 1778).

Eisinga, E., 'Beschrijving van het Planetarium', in: E. Havinga et al., *Planetariumboek*, pp. 251–330.

Ekama, C., *Oratio de Frisia ingeniorum mathematicorum inprimis fertili* (Franeker 1809).

Faber, J.A., *Drie eeuwen Friesland: Economische en sociale ontwikkelingen van 1500 tot 1800*, vols. 1–2 (Wageningen 1972).

Gothaische gelehrte Zeitungen (Gotha 1774–1804).

Halbertsma, J.H., *Hulde aan Gysbert Japiks bewezen: In de Sint Martini kerk te Bolsward, op den 7 Julij 1823*, vols. 1–2 (Bolsward 1823, 1827).

Havinga, E. et al., *Planetariumboek Eise Eisinga* (Arnhem 1928).

Jensma, G. et al., *Universiteit te Franeker, 1585–1811: Bijdragen tot de geschiedenis van de Friese Hogeschool* (Franeker 1985).

King, H.C., and J.F.M.G. Millburn, *Geared to the Stars: The Evolution of Planetariums, Orreries, and Astronomical Clocks* (Toronto 1978).

Kok, J., *Vaderlandsch woordenboek*, vols. 1–35 (Amsterdam 1780–1796).

Kuiper, J.R., *Een revolutie ontrafeld: Politiek in Friesland 1795–1798* (Franeker 2002).

Langereis, S., *Machineman: De tijden van Eise Eisinga* (Amsterdam 2024).

Lindeboom, J., 'Een patriots predikant, G.J.G. Bacot', *Nederlands archief voor kerkgeschiedenis* 38 (1952), pp. 86–117.

Martens, M. et al., *Ik heb het groote doel mijner Aardsche bestemming bereikt* (Hilversum 2005).

The Monthly Review or Literary Journal (London 1752–1789).

Mulder de Ridder, J., 'Eise Eisinga and His Planetarium', *Journal of Astronomical History and Heritage* 5, no. 1 (2002), pp. 65–87.

Nijboer, H., 'Het openen en sluiten van de markt: De economische ontwikkeling van vroegmodern Leeuwarden', in: R. Kunst et al. (ed.), *Leeuwarden 750–2000: Hoofdstad van Friesland* (Franeker 1999), pp. 159–83.

Observations sur la physique, sur l'histoire naturelle et sur les arts: Avec des planches en taille-douce dédiées a Mgr. le comte d'Artois (Paris 1773–1793).

Scheltema, J., *Eise Eisinga, de uitvinder en vervaardiger van het beroemde planetarium te Franeker* (s.l. 1818).

Telting, A., *Register van het archief van Franeker* (Franeker 1867).

Terpstra, H., *De Friesche sterrekonst: Geschiedenis van de Friese sterrenkunde en aanverwante wetenschappen door de eeuwen heen* (Franeker 1981).

Tersteeg, J., *De levensschets van Geert Reinders* (Winsum 1998).

Tigchelaar, Y.F., and E. Eisinga, *Plan tot herstel der academie te Franeker, door het provinciaal bestuur van Friesland* (Franeker 1798).

Van Albada, B.L., *Uit de oude en nieuwe doos: Herinneringen uit de school en het leven van een 80 jarigen oud-hoofdonderwijzer* (Groningen 1875).

Van der Kroe, A. et al., *Vaderlandsche letteroefeneningen* (Amsterdam 1803).

Van Heiningen, T.W., *The Correspondence of Sebald Justinus Brugmans (1763–1819)* (The Hague 2010–2011).

Van Nederveen Meerkerk, E., 'Market Wage or Discrimination? The Remuneration of Male and Female Wool Spinners in the Seventeenth-Century Dutch Republic', *The Economic History Review* 63, no. 1 (2010), pp. 165–83.

Van Nederveen Meerkerk, E., and A. Schmidt, 'Between Wage Labor and Vocation: Child Labor in Dutch Urban Industry, 1600–1800', *Journal of Social History* 41, no. 3 (2008), pp. 717–36.

Van Sluis, J., *De academie van Vriesland: Geschiedenis van de academie en het athenaeum te Franeker, 1585–1843* (Gorredijk 2015).

Van Swinden, J.H., *Beschryving van een konst-stuk, verbeeldende een volleedig bewegelyk hemels-gestel, uitgedagt en vervaardigd door Eise Eisinga* (Franeker 1780).

Van Swinden, J.H., and W. Eekhoff, *Beschrijving van het Rijks-Planetarium te Franeker, van 1773 tot 1780 uitgedacht en vervaardigd door Eise Eisinga* (Leeuwarden 1851).

Van Swinden, J.H. et al., *Beschrijving* [...] *van een kunststuk, verbeeldende een volledig bewegelijk hemels-gestel, uitgedacht en vervaardigd door Eise Eisinga* (Franeker 1824).

Vreeling, S., 'Eise Eisinga was hier: De afnemende authenticiteit van het Planetarium in Franeker rond 1900', *Bulletin KNOB* 121, no. 2 (2022), pp. 20–33.

Wassenbergh, A., 'Beroemde friesche wiskundigen', *Nieuwe Friesche volksalmanak* 2 (1854), pp. 76–96.

Zuidervaart, H.J., *Speculatie, wetenschap en vernunft: Fysica en astronomie volgens Wytze Foppes Dongjuma (1707–1778) instrumentmaker te Leeuwarden* (Leeuwarden 1995).

Zuidervaart, H.J., *Van 'konstgenoten' en hemelse fenomenen: Nederlandse sterrenkunde in de achttiende eeuw* (Rotterdam 1999).

Zuidervaart, H.J., 'De Friese "Boerenprofessor": Realiteit of mystificatie?', *De Vrije Fries* 93 (2013), pp. 99–122.

Zuidervaart, H.J., and R. van Gent, *Between Rhetoric and Reality: Instrumental Practices at the Astronomical Observatory of the Amsterdam Society 'Felix Meritis', 1786–1889* (Hilversum 2013).

Consulted Archives and Libraries

Gemeentearchief Franeker, GF
Groninger Archieven Groningen, GAG
Koninklijke Bibliotheek Den Haag, KB
Koninklijk Huisarchief Den Haag, KH
Nationaal Archief Den Haag, NA
Noord-Hollands Archief Haarlem, NHA
Tresoar Leeuwarden, Tresoar
Universiteitsbibliotheek Leiden, UBL
Universiteitsbibliotheek Universiteit van Amsterdam, UVA
University Library Yale, ULY

About the Author

Arjen Dijkstra is director of Tresoar; literature museum, archive and library of Friesland.

Brokering Knowledge

Devotees of Science

4. Electricity and Enlightenment: The Itinerant Lecturer Martin Berschitz

Oliver Hochadel

Abstract: This contribution reconstructs the life of the itinerant lecturer Martin Berschitz (ca. 1740–after 1800). Berschitz was an instrument maker and toured parts of Central and Western Europe for about three decades. His public performances focused on electricity, and he also offered to erect lightning rods. Despite his problematic status as a 'mere showman', he contributed to the circulation of practical knowledge in natural philosophy though his lectures and instruments. Berschitz was involved in numerous controversies with other practitioners and municipal authorities. The reconstruction of these skirmishes elucidates how scientific expertise and authority were negotiated in the latter part of the eighteenth century. Itinerant lecturers such as Berschitz served as a contrast-foil for the persona of a 'real' natural philosopher.

Keywords: Enlightenment, science as spectacle, lightning rod, scientific instrument

On 27 August 1792, at Kremsmünster Abbey in Upper Austria, the itinerant lecturer Martin Berschitz (dates unknown) captivated the Benedictine monks with his electrical demonstrations. In the chronicle of the monastery, Laurenz Doberschiz (1734–1799) described the performance in considerable detail. Berschitz drew sparks from the noses, tongues, and shoes of the monks. He fired a cannon remotely placed outside the 'mathematical tower' where the show took place.[1]

1 Doberschiz was 'Subprior' (deputy of the abbot), eager to justify the existence of the religious order, pointing out their scholarly traditions. At the time, Kremsmünster Abbey had quite a reputation for being a place of learning, including the natural sciences. The mathematical tower

Bijleveld, Nikolaj, Arjen Dijkstra, and Samuel Gessner, eds. *Cultivating and Communicating Natural and Technical Knowledge around 1800: Devotees of Science.*
Taylor & Francis Group, 2025
DOI 10.5117/9789463728577_CH04

Several of the demonstrations were meant to prove the protective power of the lightning rod. Others were supposed to dazzle, such as the 'killing' and subsequent 'resurrection' of a frog, or the seemingly magical embossing of a portrait of the German Emperor onto a copper plate covered by plaster, all by means of electrical discharges. The show continued after dark with a new sequence of demonstrations that were based on the effects of illumination through electricity. This included the 'most magnificent illumination of a [miniature] temple',[2] the appearance of the sequence of letters 'C.T. E.M. VIVAT' (in homage to Palatine Elector Karl Theodor (C.T.) and abbot Erenbertus Meyer (E.M.)), and the reignition of a candle that had just been doused through an electrical discharge.[3]

It is rare to come upon such a detailed description of the performance of an itinerant lecturer in the Enlightenment that even includes the reactions of the audience.[4] Doberschiz's account, six pages long, conveys both the fascination of the spectators witnessing these astonishing visual phenomena as well as the admiration they felt for the skilled performer.

This chapter reconstructs the itinerant life of Martin Berschitz as a public lecturer: his thirty years on the road and the places he performed. It describes how his portfolio of demonstrations and related services, in particular the making of scientific instruments and the erection of lightning rods, evolved. What circumstances led Berschitz to embrace this activity and what can we say about his formation and training? How did itinerant lecturers such as Berschitz fit into the panorama of natural philosophy in the late eighteenth century? To what extent did these performers participate in the circulation and production of knowledge? In what way could Berschitz claim scientific expertise and authority? How did he present himself and how was he perceived by his audiences, other practitioners of natural philosophy, and the municipal authorities?

served as an observatory and housed a collection of instruments. For the scientific culture in monasteries at the time, see Bloemer, *Empirie*, in particular chapter 4, for the Kremsmünster observatory Klamt, *Sternwarte*.

2 All translations of quotes from sources into English are mine. On occasion the original quote is in the corresponding footnote. Here: 'Illuminatio templi magnificentissimi'.

3 Stiftsbibliothek Kremsmünster CCn 350, pp. 341–47; this source (henceforth Kremsmünster) has already been quoted by Fellöcker, *Geschichte*, p. 148, n. 2, and Schrott, '"Blitzfang"', providing a partial translation into German, pp. 294–95.

4 There is a shorter and more neutral account of Berschitz's performance in the Monastery of Elchingen from December 1788, yet quite similar with respect to the demonstrations, in the manuscript of Benedikt Baader, *Merkwürdigkeiten des Reichsstiftes Elchingen von 1785–1818*, Historischer Verein für Schwaben, H 150 1/2, Stadtarchiv Augsburg, vol. 1., pp. 364–65.

In the context of this volume the persona of Berschitz provides a contrast to other practitioners of late Enlightenment natural philosophy. He was not a devotee of science like Eisinga, i.e. engaging in scientific practices *alongside* his main profession. Rather, science *was* his profession. It was precisely this feature, the fact that he pursued electricity as a business in order to earn his livelihood, that defined him in the eyes of his learned contemporaries.

Historiography and Sources

As a first step, I will briefly resume the literature on science and spectacle in the eighteenth century in order to contextualise Berschitz. 'Science and Spectacle in the Enlightenment' is by now a well-trodden subject. Historians of Enlightenment science have convincingly argued that the public performances of natural phenomena were crucial for the spread of natural philosophy. Oscillating between lecture and show, between instruction and entertainment, these experimental practices formed part of the emerging public sphere.[5] Urban centres such as Paris boasted a rich scientific culture reaching specific segments of the public marked by social status and learning.[6] Among the most popular branches of natural philosophy was the new field of electricity. Since the mid-1740s its phenomena lent themselves to spectacles full of sparks, light effects, and explosions: metallic spiders moved and bells chimed by themselves, brandy could be ignited through an icicle, the Kiss of Venus created a painful electrical discharge between the lips of man and woman, and an electrified person could attract snippets of paper.[7]

The invention of the Leyden Jar in the mid-1740s marks the beginning of this electrical show business since it allowed its users to generate very strong discharges. As a result, as the English naturalist Joseph Priestley put it, 'numbers of persons, in almost every country in Europe, got a livelihood by going about and showing it'.[8]

5 Schaffer, 'Natural Philosophy'; Bensaude-Vincent and Blondel, *Science and Spectacle*; Thébaud-Sorger, 'Spectacle de Science'.

6 Paris: Sutton, *Science*; Lynn, *Popular Science*.

7 Again, the literature has grown enormously in recent decades: England: Schaffer, 'Consuming Flame'; Maddaluno, 'Unveiling Nature'; Netherlands: Roberts, 'Science Becomes'; Germany: Hochadel, *Öffentliche Wissenschaft*; Italian context: Bertucci, 'Sparking Controversy'; North America: Delbourgo, *Most Amazing*.

8 Priestley, *History*, p. 84.

Due to the scarcity of sources, generally rather little is known about these scientific entrepreneurs. They were to a certain extent self-taught, well able to build (and repair!) scientific instruments as well as invent new ones. Most importantly, they were experienced performers, able both to skilfully manipulate their instruments and to communicate in an engaging way with their paying audiences. They often led an itinerant life, touring between cities and across national borders.

The spectrum of itinerant lecturers in the eighteenth century is quite heterogeneous. Performers such as James Ferguson (1710–1776), Benjamin Martin (1704–1782), Jakob von Bianchy (1732–1785), François Bienvenu (1758–1831), and Etienne-Gaspard Robertson (1763–1837) belong to the upper part of this spectrum.[9] They published treatises, handbooks, and even memoirs, and were well respected by the scientific community of the time (not excluding occasional skirmishes). Some of them even became members of learned societies.

Below this upper crust there were, on a European scale, dozens of far less known itinerant lecturers. That historical actors such as Martin Berschitz fell into oblivion until recently and hardly figured in the historiography of Enlightenment science is not only due to the lack of sources but also to the perception that itinerant lecturers were only business-oriented showmen, or worse, charlatans, and hence not proper subjects for the history of science.[10] This kind of boundary work, placing 'real' natural philosophers on the one side and 'mere' performers on the other, already took place in the Enlightenment itself and continued well into the twentieth century.[11]

The case of Berschitz allows for a revision of this view. As a first step the different kinds of sources that may be recovered need to be assessed. There is some evidence that Berschitz published a little treatise on lightning rods and a 'catalogue' of his machines, but they were not preserved. The main sources for reconstructing his demonstrations are his advertising material, the shorter newspaper ads and the more detailed handbill, but also some

9 These lecturers are well known: Millburn, *Benjamin Martin*; Millburn, 'James Ferguson's Lecture Tours'; Millburn and King, *Wheelwright*; on Bianchy: Hochadel, 'Business'; Tantner, 'Wiener Barometermacher'; on Robertson: Hochadel, 'Zauberhafte Aufklärung'; on Bienvenu: Bret, 'Un bateleur'; Suay Matallana and Bertomeu Sánchez, 'François Bienvenu'.

10 Schneider, *Georg Christoph Lichtenberg*, p. 37, called him (in 1955) 'a kind of charlatan, a pseudo-scientific conjurer' ('une sorte de charlatan, de prestidigitateur pseudo-scientifique'.)

11 In the German-speaking context, the work of Stephan Oettermann since the late 1970s was instrumental to recover this culture of public spectacle in the eighteenth and nineteenth centuries.

reports of eyewitnesses in newspapers and chronicles.[12] Upon arrival in a given town, Berschitz had to obtain a permit to stage his show. The municipal authorities generally kept a close watch on any kind of travelling performers and their 'morality'.[13] A considerable number of archives in half a dozen countries contain records of such requests.[14]

Berschitz also figures in letters penned by Georg Christoph Lichtenberg (1742–1799) in the aftermath of their encounter in Göttingen in early 1782. They constitute an important source for the reconstruction of Berschitz's career and persona. These represent Berschitz as seen through Lichtenberg's eyes, a fact that will have to be critically reflected. Nowadays Lichtenberg is better known as a sharp-witted writer, acclaimed for his thoughtful and often hilarious aphorisms. Yet at the time he was a professor of natural philosophy at the University of Göttingen and well known for his research in experimental physics. Lichtenberg's letters mention Berschitz and sometimes contain even quoted snippets from their mutual correspondence. This allows us to reconstruct the interaction of a highly respected member of the German community of scholars with an itinerant lecturer such as Berschitz and in particular how Lichtenberg judged him in terms of his abilities, knowledge, and ultimate motives.

I first came upon Berschitz in my research on itinerant electricians in the German Enlightenment in the late 1990s. Back then I authored a few mostly German-language publications, that dealt with Berschitz in a number of ways.[15] Since then, owing to the digitalisation of many sources, both periodicals and archival material, as well as search engines, but also thanks to the help of many colleagues (see the acknowledgements in the final footnote of this chapter), I have been able to gather a substantial number of new sources on Berschitz, in particular as regards his disputes about the proper design of lightning rods in the late 1780s and his frequent contacts with monasteries. While drawing on some of the arguments of my earlier publications, this paper provides a much fuller account of Berschitz's professional career as an itinerant lecturer.

12 Two of Berschitz's handbills can be found online: from his sojourn in Munich in 1789: https://www.germanhistory-intersections.org/de/wissen-und-bildung/ghis:document-182 and from Vienna 1796: https://www.digital.wienbibliothek.at/wbrobv/content/titleinfo/1933164, Wiener Stadt- und Landesbibliothek. Also see Hochadel, *Öffentliche Wissenschaft*, p. 197.

13 Hochadel, *Öffentliche Wissenschaft*, pp. 201–5.

14 In the following I will only partially reference the archival sources for Berschitz's sojourns. See the publications in the following note for the precise references.

15 Hochadel, '"Martinus Electrophorus"'; Hochadel, *Öffentliche Wissenschaft*, pp. 250–73; Hochadel, 'Fußtruppen'; Hochadel, 'The Sale'.

An Itinerant Lecturer on Tour

Next to nothing is known about the first decades of Martin Berschitz's life except that he was born in Vienna. The dates of both his birth and death are unknown. His name appears in the sources for the first time in April 1771 in Leipzig at the Easter Fair. From another source we may deduce that he had a son born around 1764. It is thus fair to assume that Berschitz himself was probably born in the early 1740s.

So far over fifty of his sojourns could be documented between 1771 and 1800. He visited Switzerland at least twice, first in 1775 and again in 1779, performing both times in Solothurn and Bern, and during his second trip also at the Monastery of Muri (Aargau) and possibly in Schaffhausen. Between June and November 1776, he ventured as far west as Aachen and Leuven. Sources allow us to reconstruct one large journey that lasted from late 1781 until possibly 1784. The first stretch of this journey (1781–1782) led Berschitz to central and northern Germany (Kassel, Göttingen, Hanover, possibly Hamburg, Bremen, and Münster) before he ventured for a second time to the north-west. He visited Amsterdam in the Dutch Republic (March and April 1783), Ghent, Mechelen, and possibly Brussels in the Austrian Netherlands and Lille in northern France (in the second half of 1783). In 1784 Berschitz performed in Aachen and Hanau, apparently 'returning' to the south-east of the German Empire.

Most of his sojourns throughout the first two decades (1770s and 1780s) took place in what is today Bavaria. In Augsburg, for example, he set up camp at least four times (1773, 1778, 1789, 1791). One possible reason (although the evidence is indirect) for this geographical focus was the suppression of the Society of Jesus in 1773. This posed a challenge to the education system in this predominantly Catholic part of the German Empire. Many institutions of learning were not only in need of new teachers but also tried to acquire scientific instruments and equipment in order to teach natural philosophy. In 1778, Berschitz sold electrical instruments to the St Salvator School in Augsburg, a former Jesuit Gymnasium. While in Amberg in 1790, Berschitz claimed to have received 'permission from the higher state government Munich' to show his 'mechanical arts' in the territories of Bavaria and Oberpfalz and to have visited previously Mannheim, Heidelberg, Zweibrücken, Ingolstadt, Munich, Landshut, Straubing, Neuburg an der Donau, 'and other places, where there are public schools'. There he 'restored the cabinets [of instruments] and instructed the teachers'.[16] With the exception

16 Request Berschitz to Magistrate, 1 February 1790; Staatsarchiv Amberg, Fürstentum Obere Pfalz, Regierung Amtsbücher und Akten 1053, 'von der oberen Landes Regierung München die

of Zweibrücken, there were former Jesuit schools in all these towns. And by that time, they were all part of the state of Kurpfalz-Bayern. In 1777, the Kurfürstentum Bayern (Electorate of Bavaria) and the Kurpfalz (Electoral Palatinate) fused to become Kurpfalz-Bayern ruled by the Palatine Elector Karl Theodor (1742–1799). Berschitz had paid homage to this high-ranking prince from the House of Wittelsbach in his performance in Kremsmünster in 1792, mentioned above, and referred to himself at times as 'Imperial and Palatine Court Mechanic'.[17] The fact that Berschitz apparently had some kind of endorsement from the Palatine court may explain why he specifically targeted schools in Kurpfalz-Bayern that needed instruments and instructions how to use them.

The revolutionary wars in the west from 1792 onwards might have restricted his movements. This would also explain why all of his documented sojourns between 1793 and 1797 (again, the record is patchy) took place in Vienna, Bratislava (Pressburg), and Bohemia (Prague, Plzeň /Pilsen, Klatovy/ Klattau, and Pisek).

Berschitz did not travel alone but with two other persons. One chronicle mentions that his two helpers included his son Joseph, who was around 18 years old in 1782. According to a source from 1788, Berschitz travelled with his wife 'and a young cousin'.[18] Just a few months later he told the Magistrate of the City of Ulm that he had to pay board and lodging for three persons, hence pleading for an increase in his honorarium.[19] An itinerant lecturer needed helpers: for assistance during the shows but also concerning logistical issues such as the transport and maintenance of the instruments.

Berschitz's itinerary has some considerable gaps of several years (e.g. between December 1779 and December 1781, September 1784 and July 1786, and between 1797 and 1800) where no activity is recorded. We might therefore assume that the over fifty documented sojourns only represent a part of his itinerary during these three decades. Another (complementary) explanation for the lacunae in the itinerary might be that Berschitz had some kind of 'home base' in between journeys. He might have disposed of a permanent workshop to make instruments and to service customers in the relative vicinity with electrical machines and lightning rods. There are

gnädigste Licenz erhalten habe meine Mechanische Künste', 'und anderen orthen, wo offentliche Schullen synt, [...] die Armaria hergestellet – und die Professores unterrichtet habe'.

17 *Augsburger Intelligenz-Zettel*, Num. 2 – 12.1. 1778, 'Kaiserl. und Churpfälzischer Hofmechanikus'.

18 Chronik von Kloster Fürstenfeld aus der Chronik von Abt Gerard Führer, § 326, p. 259, April 1788, Bayerische Staatsbibliothek, Hss Cgm 3920 (henceforth Fürstenfeld Führer).

19 Request Berschitz to City Council, 14 November 1788, A [1619], no. 3, Stadtarchiv Ulm.

some hints in the sources, far from conclusive, that this base might have been located in Munich between 1786 and 1788 and maybe in Prague around 1795/96.[20] Vienna as a possible base also comes to mind, but Berschitz seems to have hardly performed in his home town. For this city, only a series of performances between September and November of 1796 have been documented.[21]

The length of his stays could differ substantially and varied between a few days and several weeks. Lasting over two months, Berschitz's sojourn in Vienna was by far the longest. The decisive factor was the (continuing) demand for his services, performances, and instruments. As mentioned above, Berschitz usually had to obtain a permit from the local authorities upon arrival, which was mostly but not always granted, as well as to find a suitable venue for his performance. Very often he would rent the back room of an inn, and on occasion also a theatre. In general, he would perform once a day (sometimes twice) for two hours, usually in the late afternoon or early evening. The admission fee was typically divided in two categories (sometimes three): first and second seats, priced at twenty and ten *kreuzer* respectively, while persons of rank (*Standespersonen*) paid 'at will'. On occasion the handbill specified that no more than thirty spectators would be admitted, to guarantee a good view of the show.

In July 1800, Berschitz announced the end of his travels in the *Wiener Zeitung*. Yet he continued to offer his portfolio of services within the city as the advertisement explained in great detail.[22] At the address given, his son Joseph was listed as a 'mechanical instrument maker' in an official registry the same year.[23] I could not find any more sources on Martin Berschitz or his son dated after 1800, so the dates of their deaths are unknown.

Berschitz's Performances: An Evolving Portfolio

This section tries to reconstruct what Berschitz actually showed in his performances and how these demonstrations evolved between 1771 and 1800. In his first documented sojourn at Leipzig in April 1771 Berschitz

20 For Prague: Urbanus Leopoldus Rosche, Annales Strahovienses (1791–1803), p. 170, Strahov Library Prague, loc. no. DI IV 5'; Pařez, 'Strahov', p. 46, n. 19.

21 There are six virtually identical announcements between early September and mid-November. 'Physikalische Experimente'. *Wiener Zeitung*, 7 September 1796, pp. 16–17 (pp. 2590–91), 16 November 1796, p. 15 (p. 3277).

22 *Wiener Zeitung*, 19 July 1800, p. 2330.

23 *Hof- und Staats-Schematismus*, p. 128.

demonstrated an 'optical machine and different tricks'. By the end of the 1770s his programme seems to have switched to displaying different phenomena of electricity, although not exclusively. His advertisements and also some of the reports mention 'machines' and the effects of magnetism. Among the dozen or so itinerant lecturers who operated in the German Reich in the second half of the eighteenth century, Berschitz seems to be the one most devoted to electricity.[24] The demonstrations of the others were centred on mechanics (machines), optics, acoustics, and other fields of natural philosophy. One performer who also focused very much on electricity was Johann Conrad Gütle (1747–1827). Yet he was on the road for only about seven years before settling down in Nuremberg in 1788 where he continued to build instruments and teach natural philosophy, among other activities.[25]

Doberschiz's description of the abovementioned Kremsmünster performance coincides largely with the content of the handbill of the late 1780s. Apart from the demonstrations mentioned at the beginning of this chapter, this included the igniting and burning of sponges attached to human bodies and the melting of a sword in its sheath without damaging the latter.

Showing the electrical nature of lightning and demonstrating the protective quality of the lightning rod had a prominent place in Berschitz's repertoire in the 1780s and 1790s. One of the most popular demonstrations of electrical shows in the late eighteenth century was an experiment with a miniature house or church, traditionally called a thunder-house.[26] By means of a Leyden Jar, the model building was exposed to a strong electric discharge presented as lightning. Depending on whether the thunder-house was equipped with a miniature lightning rod or not, it would either be left unharmed or collapse or even go up in flames if an inflammable substance had been put inside (Illustration 5).

In the description of the show at Kremsmünster, Doberschiz depicted Berschitz's double demonstration of how the 'lighting touched the house, once with and once without fire'. The chronicler also added a seemingly picturesque detail: two painted figures of peasants were knocked over by the simulation of the lightning strike.[27] This might be read as a reference to the alleged resistance against the introduction of the lightning rod among parts

24 For an analysis of the range of itinerant lecturers in the German Empire see, Hochadel, *Öffentliche Wissenschaft*, chapter 4.1.

25 Rüger, 'Populäre Naturwissenschaft'; Kett, 'Er brachte'.

26 Hackmann, 'Lightning Rod'. Cf. the contribution by Nikolaj Bijleveld to this volume.

27 'Domum [...] quomodo fulmen hanc tangat cum, vel sine incendio', Kremsmünster, p. 342, § 188.

Illustration 5. The thunder-house demonstration was very popular at the time and may still be on display in technology museums today. Illustration from D. Beck, *Kurzer Entwurf der Lehre von der Elektricität. Verfaßt zum Gebrauche seiner Zuhörer* (Salzburg 1787) (*Collection Bakken Museum/ Minneapolis*)

of the population. As enlightened rhetoric would have it, common people, particularly in the countryside, were trapped by erroneous conceptions and deep-rooted superstition.[28]

While in Regensburg in October 1786, Berschitz published the advertisement for his show in the local weekly, as was his custom. Yet in the second week (i.e. after one week of performing) he added a remarkable phrase to the advertisement. He now specifically addressed 'common and unlearned people' and announced that he would 'disprove prejudices sensually' through the thunder-house demonstration.[29] Berschitz deliberately used these tropes of enlightened discourse to fight against supposedly backward understandings of the lighting rod. This shows that he was fully aware of (and arguably identified with) the current rhetoric in favour of the lightning rod. More than

28 In fact, it is very hard to find any solid evidence of people resisting the introduction of lightning rods; Dross, 'Gottes elektrischer'; Hochadel, '"In nebula"', p. 53; Schock, 'Donnerstrahl'; Bloemer, *Empirie*, p. 174.

29 *Regensburgisches Diarium oder wöchentliche Frag- und Anzeige-Nachrichten*, 24 October 1786 (no. 43), p. 342.

that, this phrase might be read as a message to the *Aufklärer* (proponents of enlightenment) that he was one of them.

This pledge to enlightened values did not stop him from cheating in order to make his performance even more impressive. At the beginning of this chapter, we mentioned the trick in which a portrait of the German Emperor appeared on a plate of plaster. Berschitz pretended that this image was created by means of electrical discharges, when in fact a copper plate with the portrait already engraved had only been covered underneath a thin layer of plaster. This trick was a regular feature of his performance as he announced it on his handbill. While the monks in Kremsmünster were apparently easily duped, Lichtenberg understood how the trick worked. Lichtenberg claimed that Berschitz had more 'scams' at the ready, yet that he would not dare to show them to him.[30] Berschitz seemed to have a good instinct to tailor his demonstrations to the varying capacities of his audiences to discern an authentic effect from a ruse. In other words he knew what to show and to whom without being exposed.

Offering dramatic demonstrations and pursuing an Enlightenment agenda were no contradiction in Berschitz's performance. In his handbill he also mentioned underwater explosions which were an optional part of his show, 'on demand', so to speak. Some local authorities did not allow these apparently dangerous experiments, as seems to have been the case during his sojourn in Bern in December 1779.[31] Years later, in 1792, Doberschiz was relieved that nobody got hurt during 'so many dangerous experiments' with considerable amounts of gunpowder.[32]

At least on two occasions, in Kassel in December 1781 and Bremen in July 1782, Berschitz was able to stage outdoor performances. These included explosions on a large scale and were thus quite different to the parlour demonstrations electrical performers used to offer. The basic principle was to ignite gunpowder over a considerable distance through an electrical discharge. To add to the drama the objects to be blown up were small ships placed on water or a model of a powder tower near the edge of the water.[33]

The pedagogic message of the explosion orgy was once more to 'show the speed of lightning'. Ships (often carrying gunpowder) and in particular powder magazines were among the objects considered in the greatest danger

30 Lichtenberg to Reimarus, 2 May 1782; Joost and Schöne, *Lichtenberg, Briefwechsel*, p. 318, 'Betrügereyen'.

31 Ratsmanual, vol. 350, 16 December 1779, p. 321, Stadtarchiv Bern.

32 Kremsmünster, § 189, p. 344, 'experimentis tot periculosis'.

33 *Augsburgische Stats- und Gelehrte Zeitung*, 16 January 1782, p. 44.

during a thunderstorm. When lightning struck the church tower of the San Nazaro Church in Brescia on 18 August 1769, it led to the explosion of the nearby powder magazine. The destruction caused was apocalyptic. Parts of the city were destroyed, and the death toll given varied between several hundred and several thousand people. This tragedy was widely reported and alerted governments, the military, and naturalists all over Europe to the impending danger of unprotected powder magazines, speeding up the installation of lightning rods.[34]

In Bremen, Berschitz carried out his explosive performance twice on the river Weser in August 1782. According to a newspaper report, the event drew 70,000 spectators on both occasions. This is certainly an exaggeration but one may assume that the audience was indeed large.[35] The performance must have been so impressive that the preacher of the dome (*Domprediger*) in Bremen, Heinrich Heeren (1728–1811), mentioned it in a sermon the following year – as proof of man's power over electricity and thus also of the effectiveness of the lightning rod.[36] The clergy was often an important ally of the propagandists of the lightning rod.

The outdoor performance in Kassel in December of the previous year had also been a huge personal success for Berschitz. A newspaper report highlights the presence of the Landgrave of Hesse-Cassel, Frederick II (1720–1785), and other persons of high rank.[37] To conduct demonstrations in front of a landgrave was every itinerant lecturer's dream. An aristocratic audience ennobled the performance and yielded valuable social prestige. The landgrave seems to have been so impressed that he ordered an electrical machine and other instruments for 300 guilders.[38] On 12 January 1782, Berschitz could even pen his name in the book of visitors of the Cabinet of Frederick II, about a month after his explosive performance, possibly when he delivered the instruments (Illustration 6). This occasion might have included another, more intimate performance.[39]

34 Bertholon, *De l'électricité*, p. 268; Hochadel, *Öffentliche Wissenschaft*, p. 143; Home, 'Points or Knobs', p. 101.

35 *Augsburgische Ordinari Postzeitung*, 7 August 1782. Reprinted in Buchner, *Das Neueste*, vol. 3, p. 313. There is a shorter version of the report in *Augsburgische Stats- und Gelehrte Zeitung*, 7 August 1782, p. 608.

36 Heeren, *Ueber die Verehrung*, p. 13. See the chapter by Bijleveld in this volume.

37 *Augsburgische Stats- und Gelehrte Zeitung* (Augsburg), 16 January 1782, p. 44.

38 The price is mentioned in a letter from Georg Forster to Anonymous, 20 November 1782; Stargardt, *Autographen*, p. 154.

39 Besucherbuch von Kunsthaus und Museum Fridericianum, Universitätsbibliothek Kassel, Handschriftenabteilung 20 Ms. Hass. 471, p. 171, 12 January 1782; Linnebach, *Das Museum*, p. 39.

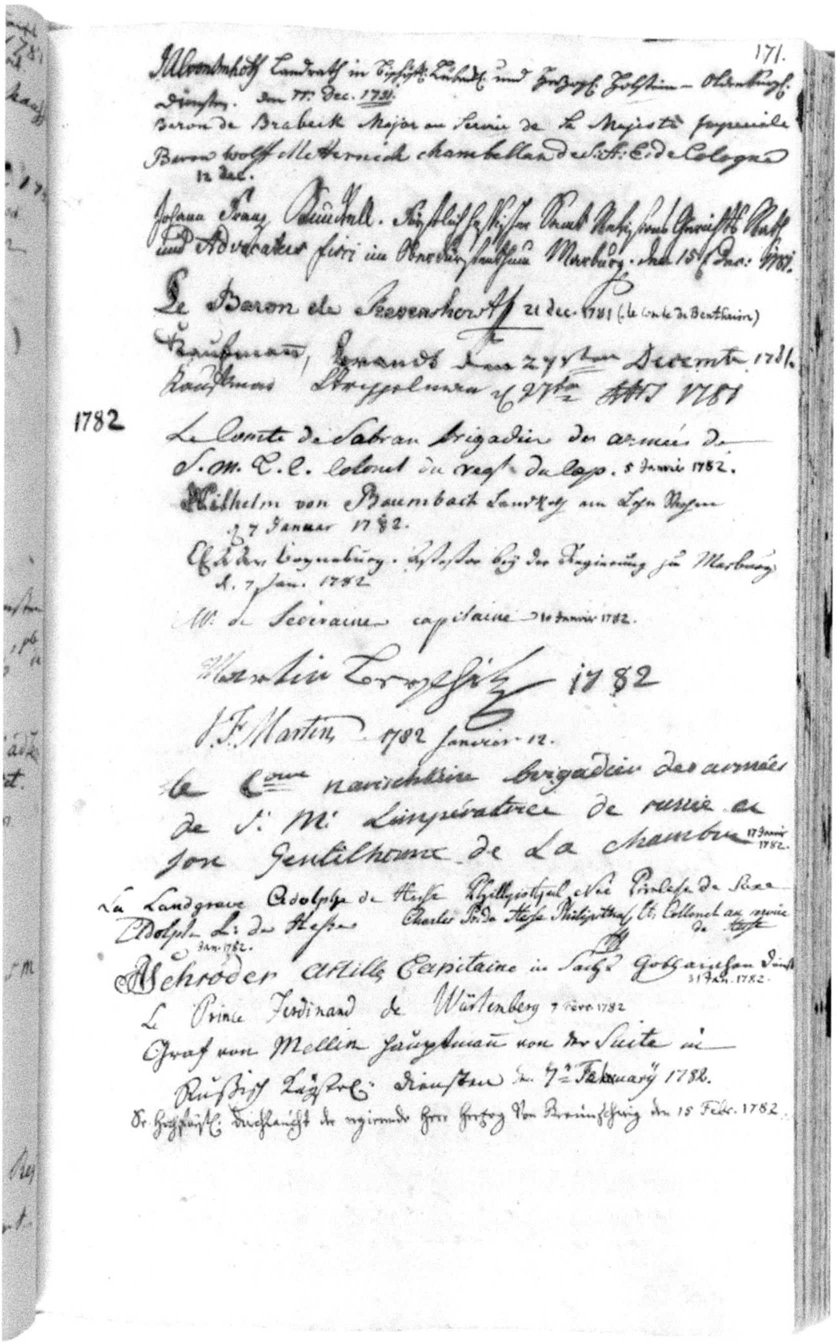

Illustration 6. Berschitz knew how to write despite the malicious comments of his contemporaries. From the book of visitors of the Landgrave of Hesse-Cassel, 12 January 1782 (*Collection Universitätsbibliothek Kassel Handschriftenabteilung, Ms.Hass. 471)*

Quite a number of advertisements for itinerant lecturers appear in newspapers published in the last decades of the eighteenth century. Yet it is rather rare to encounter an observer's report of the performance. That both performances, in Kassel and Bremen, received such enthusiastic coverage indicates that Berschitz achieved something out of the ordinary. In the early 1780s he was at the height of his fame. In 1783 he was even mentioned in the article on electricity in the *Deutsche Encyclopädie*: 'Some of these itinerant electricians proved to be very successful, such as, for example, a certain Martin Berschitz. He shows in our current times the most striking, extraordinary and powerful experiments'.[40]

Itinerant lecturers had to be flexible in several respects. Would they be allowed to stage particular experiments or not? Berschitz's open-air demonstrations were a point in case. Could they capitalise on new demonstrations that had become fashionable? This kind of flexibility required a serious effort from the itinerant lecturer as new knowledge, materials (in a broad sense), and skills were required. Three episodes from the early 1780s serve to illustrate the complexity for Berschitz to devise new attractions: gas chemistry, the automaton, and the balloon.

Example 1

Since the mid-1770s 'inflammable air' (later on called hydrogen) and 'dephlogisticated air' (oxygen) were major topics of discussion among European natural philosophers. A crucial practical issue was the question of where to collect, and respectively how to produce these gases. Scientific lecturers such as Berschitz were naturally interested in expanding their repertoire of demonstrations: with both kinds of gases, respectively the proper mix of both, ear-deafening bangs and beautiful light effects could be produced. According to a newspaper report Berschitz already demonstrated in Kassel in December 1781 'entirely new experiments of different kinds of inflammable air'. Lichtenberg was very much interested in this issue, too. He discussed the matter of the new gases at length with Berschitz when they experimented on it together. According to Lichtenberg, he had to teach Berschitz 'how to produce dephlogisticated air'. After he had left Göttingen, Berschitz kept sending Lichtenberg letters inquiring about the exact procedure. '[H]e must have completely forgotten it', surmised the physicist. 'He knew well that his audience would have been much larger had he had this air'.[41]

About half a year later, in October 1782, while in Münster, Berschitz asked Lichtenberg to share the details of 'the experiment with the watch spring',

40 Müller, 'Electricität', p. 215.

41 Lichtenberg to Wolff, 10 June 1782; Joost and Schöne, *Lichtenberg, Briefwechsel*, p. 344.

which required dephlogisticated air. Lichtenberg described his letter as 'very pathetic'.[42] Lichtenberg was probably unaware that Berschitz had fallen into serious trouble in Münster. Because he had refused to pay the rent for the local theatre where he had staged his show, the magistrate confiscated his instruments as a security deposit. Berschitz was then probably desperate to learn about the new experiment that had become very popular. Combusting a watch spring provided a very bright glow and thus promised paying audiences. Lichtenberg would not help him out, though, because, as we will see below, Berschitz had improperly used his letter of recommendation. In the long run, the itinerant lecturer got along without the advice of the professor. According to his handbill, Berschitz had incorporated a number of experiments with inflammable air into his repertoire by the mid-1780s. In 1788, in the monastery of Elchingen, he even showed the monks how to produce 'inflammable air'.[43]

Example 2

While in Ghent in August 1783, Berschitz announced in the local newspaper that he would display a kind of chess automaton: 'qui joue aussi naturellement, qu'une autre figure, qui à Vienne joue aux échecs'.[44] He was alluding to the chess automaton of the engineer Wolfgang von Kempelen (1734–1804), who constructed the 'famous Turk' around 1769, a machine that was supposedly able to play chess.[45] Von Kempelen displayed his machine first at the Habsburg Court, but only occasionally later on. In 1783/84 he toured Western Europe. In April 1783 he went to Versailles and in May he came to Paris.[46] Berschitz was clearly aware of the sensation that the 'famous Turk' had caused in the French capital and hoped to attract an audience eager to see something similar.

Example 3

The advent of ballooning in the summer of 1783 was a sensation on a scale hard to imagine nowadays. Each ascent into the sky drew thousands of spectators.[47] Berschitz and quite a few other itinerant lecturers ventured

42 Lichtenberg to Schernhagen, 17 October 1782; Joost and Schöne, *Lichtenberg, Briefwechsel*, p. 448.

43 Baader, *Merkwürdigkeiten*, pp. 364–65.

44 *Gazette van Gend*, 21 August 1783, p. 519.

45 Felderer and Strouhal, *Kempelen*.

46 Standage, *The Turk*, pp. 44–52.

47 The literature on ballooning in late eighteenth-century Europe is considerable, see e.g. Thébaud-Sorger, *L'Aérostation*; Lynn, *The Sublime Invention*; Kim, *Imagined Empire*.

into the balloon business. The business model of this kind of public spectacle was quite different to an evening lecture, as the construction of the balloon was generally paid by subscription. Already in September 1783, just a few weeks after he attempted to reproduce a chess automaton, and less than a hundred kilometres to the south-west of Ghent, Berschitz tried to launch an air balloon in Lille. The attempt failed and those who had financed the balloon were very aggravated. On 14 December 1783, Berschitz announced another launch in Mechelen, and in April 1784 in Aachen, but it is unclear whether these attempts were successful (Mechelen) or ever took place (Aachen).[48]

Two of the three ventures summarised here remained brief episodes in Berschitz's career. Apparently, they did not prove profitable (the chess automaton) or even feasible (the balloon). Despite some problems he was able to master the demonstrations based on the new gas chemistry. These three examples show that he tried to adapt rapidly to new techno-scientific advances or inventions catering to the taste of his audiences.

Due to the scarcity of direct testimonies, it is generally difficult to reconstruct how these audiences reacted toward the performances of itinerant lecturers. Yet, as shown above, in the case of Berschitz's performances, there are several sources such as the two newspaper reports from Kassel and Bremen as well as pastor Heeren's report that do convey the excitement and delight of his spectators. As already pointed out at the beginning of this chapter, Doberschiz too was deeply impressed by Berschitz's sensually overwhelming demonstrations: 'I have seen a few things in the last two hours of this art ('arte') that neither I nor any *Cremisanus* [monk of Kremsmünster] had ever seen before'.[49]

Other eyewitnesses referred less to the amazement they experienced than to the dexterity of the performer. Lichtenberg granted Berschitz that he was a manually skilled experimenter and acknowledged that he did some 'pretty splendid experiments' and that his 'melts are also very nice, yet he made a bit too much noise about it'.[50]

Franz Ferdinand Wolff, a lawyer and amateur scientist, who had attended *all* of Berschitz's performances in Hanover in May 1782, was still full of respect for his experimental skills twenty years later. 'I never noticed in 14

48 Authorisation for Berschitz, 24 September 1783, A 6 703, police des spectacles (registre n°65, fol. 13); reaction of public: C 703 O 10, Archives municipales de Lille; *Stadt-Aachener Zeitung*, 24 April 1784, according to Pauls, 'Der Luftschiffer', p. 54; Thébaud-Sorger, *L'Aérostation*, p. 168.
49 Kremsmünster, § 188, p. 344.
50 Lichtenberg to Johann Daniel Ramberg, 8 April 1782; Joost and Schöne, *Lichtenberg, Briefwechsel*, p. 303.

evenings that he had failed even a single demonstration'.[51] The fact that Wolff visited so many performances testifies to his fascination or even admiration for Berschitz. By contrast, in the correspondence with his 'superior in natural philosophy' Lichtenberg, Wolff would emphasise 'that all his experiments were performed by me long before, as Bergschütz [Berschitz] has to admit himself'. The above-quoted encyclopaedia article of 1783 used a very similar kind of rhetoric: Berschitz's experiments were 'well known to all physicists' and he showed them 'for money'.[52] In short: he was an entrepreneur of science, not an innovative researcher.

And just like Lichtenberg, Wolff tried to ridicule Berschitz: 'On his departure Bergschütz has left enchanted people behind him: at all corners and places one can see glasses being turned untiringly by all kinds of people'.[53] We might also read Wolff's jibe against the grain, namely as a confirmation that Berschitz was indeed able to generate interest in electricity among his audiences, animating them to experiment themselves.

Lecturer, Healer, and Instrument Maker

Berschitz's performances were the centrepiece of his itinerant career, yet he also offered a number of other services and commodities. It might be most helpful to think of Berschitz's professional output in terms of an evolving and changing portfolio that also included instruction, electrotherapy, the sale of scientific instruments, and the installation of lightning rods.

Berschitz clearly saw himself as more than a mere performer. The category of itinerant *lecturers* conveys this explicitly. It was part of their self-understanding that they taught scientific knowledge *through* demonstrations. Unlike the 'non-itinerant' lecturers in urban centres such as London or Paris, who offered courses that lasted for several weeks or even months, Berschitz would only offer one class.[54] As we have seen, his performance consisted of a sequence of spectacular demonstrations rather than of verbal explanations.

On occasion Berschitz would even become a sort of private tutor, paid to teach specifics to his customers. There are only a few examples to be found

51 Wolff, 'Etwas über Blitzableiter', p. 70, n.

52 Müller, 'Electricität', p. 215.

53 Wolff to Lichtenberg, 17 May 1782; Joost and Schöne, *Lichtenberg, Briefwechsel*, p. 331.

54 Examples: Paris: Lynn, *Popular Science*, chapter 3; London: Millburn, 'The London Evening'.

Illustration 7. Still on display in the Strahov Monastery in Prague: an electrical machine sold by Berschitz in March 1795. Photo by Jan Pařez (*Collection Strahov Monastery*)

in the sources, but it is telling that they relate to 'technical' aspects rather than theoretical ones. That is to say, he was asked to divulge how to build the instruments and in particular how to use them but was not expected to elaborate on the explanation of, e.g. some electrical phenomenon. This is illustrated in the correspondence of the St Salvator School in Augsburg on the occasion of the purchase of electrical instruments in January 1778.[55] In 1779, Berschitz not only sold the monks of the Benedictine Monastery of Muri a set of instruments, but also charged another 25 guilders for 'teaching', presumably how to handle the instruments.[56] In 1796, the Strahov Premonstratensian Monastery in Prague (which had purchased an electrical machine from Berschitz the year before: Illustration 7) hired Berschitz to teach the librarian Kašpar Boušek (1757–1828) how to use a set of recently acquired instruments.[57]

Like other itinerant electricians, Martin Berschitz also practised electrotherapy, at least occasionally.[58] While at the monastery at Fürstenfeld, on

55 Request St. Salvator School to City Council, 31 January 1778, Handwerkerakten Kt. 2, Fasz. 3, Baumeisteramt, Erfindungen, Stadtarchiv Augsburg.

56 Baumer-Müller, 'Die Gebrüder', p. 355, n. 42.

57 Haubelt, 'Vznik Prirodovedeckych', pp. 79–80; Pařez, 'Strahov', p. 46.

58 Gütle, for example; Hochadel, '"My Patient"', pp. 79–81; Kett, 'Er brachte', p. 190.

30 April 1788, he treated abbot Tecelin who suffered from bad hearing, yet without success, as the chronicler remarked.[59] Already since mid-century electrical practitioners had tried to cure all kinds of illnesses by applying electricity to the human body. It was very much a trial-and-error procedure.[60] For electrical performers electrotherapy was a tricky business because they entered the much more regulated field of medicine (as opposed to natural philosophy). Berschitz did not offer electrotherapy in the advertisements of his shows during the period of his itinerant career. Only when he announced that he was settling down in Vienna in 1800 did he offer to treat 'the sick in need of electricity', not without specifying that this would happen only 'on the orders of the doctors'.[61]

Many of the itinerant lecturers offered scientific instruments for sale. It is difficult to say how they were actually able to manufacture them while on the road. As they had to travel light, we might assume that they built these scientific wares on demand and *in situ*. Given the lack of a workshop of their own and the basic materials, they had to collaborate with local craftsmen. While in Augsburg in October 1791, Berschitz let the city council know that he received commissions 'from nearby monasteries and other respected gentlemen' and that therefore 'the work of local professionals and craftsmen' was 'indispensable' for him.[62] Particularly in the case of electricity (but also optics) the practical problems of the itinerant lecturers should not be underestimated: friction bodies made of glass, the core of an electrical machine, were not only very fragile (they might break on the bumpy roads or during the show itself), but also difficult to obtain.[63]

Monasteries were among his best clients: they often bought entire sets of instruments.[64] In 1779, Berschitz sold instruments for 300 guilders to the Swiss Benedictine Monastery of Muri.[65] Around 1790 he seems to have been particularly successful marketing his instrument to monasteries in Bavaria and Upper Austria: in April 1788 he sold an electrical machine with many accessories (no price given) to the Monastery Fürstenfeld; in December 1788 a set of instruments to the Monastery Elchingen for 300 guilders; in August 1792 an electrical machine, 'accessories', and 'experiments' to the

59 Fürstenfeld Führer, § 326, p. 259; see already Bloemer, *Empirie*, p. 191.

60 Hochadel, '"My Patient"'; Zanetti, 'Curing with Machines'.

61 *Wiener Zeitung*, 19 July 1800, p. 2330.

62 Request Berschitz, October 1791, Theater 23/15, Stadtarchiv Augsburg.

63 Hochadel, *Öffentliche Wissenschaft*, pp. 201, 226–27.

64 See already Schrott, '"Blitzfang"', p. 294.

65 Baumer-Müller, 'Die Gebrüder', p. 355, n. 42.

Monastery Kremsmünster, for 300 guilders;[66] and around the same time (no exact date of sale or price known) 'mathematical-physical instruments' to the Cistercian monastery in Waldsassen.[67] In March 1795 Berschitz sold 'electrical appliances', including an electrical machine, for 150 guilders to the Strahov Monastery in Prague.[68]

In most of the cases just mentioned the sales of his electrical instruments had been preceded by well-received performances. This indicates that shows may be considered the most convincing part of his publicity. The fact that at least on three occasions Berschitz sold electrical instruments for 300 guilders might indicate that he offered a 'package' that appealed to his better-off clients, such as the Landgrave of HesseCassel who paid the same price. Monasteries seemed to have the necessary means to purchase electrical machines plus accessories. They often already had collections of instruments into which Berschitz's products could be integrated. Secondary schools eager to offer instruction in the 'new natural philosophy', such as the numerous former Jesuit schools in Bavaria, were also an important group of customers.

Berschitz offered to construct electrical machines of three different types that were in use at the time, defined by the shape of the friction body: plates, cylinders, or globes.[69] And in Vienna in November 1796, he offered a 'booklet' that explained all the machines he sold and the experiments one could perform with them.[70] This kind of commercial publication was typical for instrument makers of the time.

While most of Berschitz's instruments are lost, a few have been preserved. There is a Leyden Jar at the Cabinet of Astronomy and Physics in Kassel. 'MARTIN BERSCHITZ MECHANIC D'WIEN' is written across it. The electrical machine the landgrave purchased in December 1781 was still listed in 1814 but is probably not among the five historic machines the cabinet holds today.[71] The electrical machine Berschitz sold to the Strahov Monastery in 1795 is still kept in the museum although it has been painted bright red.[72]

66 Fürstenfeld Führer, § 326, p. 259; for Elchingen, see Baader, 'Merkwürdigkeiten', p. 365; Kramer, 'Wissenschaft', p. 282; Kremsmünster, p. 347, § 191.

67 Binhack, *Geschichte*, p. 157.

68 Haubelt, 'Vznik Prirodovedeckych', p. 80.

69 Handbill Munich, see n. 12.

70 'Nachricht', *Wiener Zeitung*, 30 November 1796, p. 19 (p. 3421), 'Büchlein'. Other (itinerant) lecturers published similar books, combining instruction with a catalogue of their instruments for sale, see e.g. Gütle, *Beschreibung*.

71 Cabinet of Astronomy and Physics, Kassel. Coester and Gerland, *Beschreibung*, p. 44; Auskunft Karsten Gaulke, Astronomisch-Physikalisches Kabinet, 16 February 2023.

72 Pařez, 'Strahov', pp. 46–47; on the restoration of Berschitz's electrical machine, see Mergl, 'Restaurovaná válcová'.

The Formation of an Itinerant Lecturer: Skills and Knowledge

In order to better situate Berschitz in the context of Enlightenment science, I will now look into how he acquired his practical and theoretical knowledge in instrument making and natural philosophy. What is known about his training? How did he keep on top of developments in the research on electricity? Which other relevant actors did he know? Given the scarcity of sources, his own formation remains difficult to reconstruct. Little is known about any schooling of the 'mechanic' Berschitz. According to one source 'he was a servant to an Italian physicist for a while; that's where he learnt his skills'.[73] This might refer to the afore-mentioned Jakob von Bianchy (Giacomo Bianchi) who had come to Vienna in the early 1760s and taught courses in experimental physics. Berschitz must have been capable of working with metal, wood, and glass. One may speculate that he went through a kind of apprenticeship in a profession that provided some of these abilities. Possibly he went on tour as a journeyman, acquiring more skills in the workshops of instrument makers.[74]

Berschitz told Lichtenberg that he had been a handyman ('Handlanger') to Joseph Franz (1704–1776).[75] Franz was one of the most prominent Jesuit scholars at the time: professor of experimental physics at the University of Vienna since the mid-1740s and tutor of the children of Empress Maria Theresa (including the later Emperor Joseph II). He was well known for demonstrating electrical and other natural phenomena to students and visitors.[76] Franz also staged public performances in the presence of the Imperial family in Vienna.[77] Yet in the 1740s, Berschitz was at best a child. It is conceivable that Berschitz assisted Franz during his lectures at the University of Vienna in the late 1750s or 1760s. Curiously, Doberschiz had been a student of Franz's physics class in the late 1750s.[78] Franz's lectures were described as giving priority to visual and spectacular phenomena over

73 Georg Forster to Anonymous, 20 November 1782; Stargardt, *Autographen*, p. 154. 'Er ist bey einem italienischen Physiker eine Zeitlang Bedienter gewesen; da hat er denn auch seine Künste her'.

74 Kett, 'Er brachte', p. 191, makes a similar case for Gütle.

75 Lichtenberg to Schernhagen, 17 October 1782; Joost and Schöne, *Lichtenberg, Briefwechsel*, p. 448.

76 Hellyer, *Catholic Physics*, p. 189 (quote); Lackner, *Die Jesuitenprofessoren*, vol. 1, pp. 152–55; Pärr, *Maximilian Hell*, pp. 15–16, 50, 61–62.

77 *Wiener Diarium*, Nr. 43, 8 June 1746; for other public demonstrations of Franz see *Wiener Diarium*, Nr. 24, 23 March 1748; Nr. 102, 21 December 1757; Nr. 30, 14 April 1759; Nr. 42, Anhang, 26 May 1762.

78 Pärr, *Maximilian Hell*, p. 168.

theoretical explanations. Although the settings for Berschitz's performances were quite different from Franz's, their underlying approach was very similar. This fact could suggest that Berschitz apprenticed with the Jesuit polymath, and even more so as he used a similar long-distance transmission for his underwater explosions. Berschitz correctly assumed that Lichtenberg knew Franz, but Berschitz's mentioning Franz failed to impress the Göttingen professor. In a mixture of condescension and commiseration, Lichtenberg states: 'All in all, he is a guy who is to be pitied that he was not taught in good time'.[79]

The learned monks who got to know Berschitz judged in a similar way. Cistercians, Benedictines, and Premonstratensians were very sensitive as regards his lack of academic training. For them, he was a 'mechanic – but only mere mechanic; for he was ignorant even of writing' (Monastery Fürstenfeld 1788), a 'man who had not studied any letters' (Monastery Kremsmünster 1792), or a 'man without any study of physics' (Monastery Strahov 1795).[80]

Skilled and dexterous with his hands but without any 'proper' knowledge acquired in acknowledged institutions, this was a typical formula by which the learned world of the Enlightenment judged Berschitz and other itinerant lecturers.[81] Historians of science have fundamentally questioned this devaluation of manual skills and have described how instrument making and other artisanal practices were based on an epistemology of its own and how this 'embodied' knowledge was crucial for scientific progress.[82]

At the same time, it is intriguing to ask what Berschitz might have read and how he might have gathered information related to his performances. In his handbill, he announced that he 'will clearly present the famous Mr Franklin's theory of electricity by means of a special machine and experiments'.[83] And according to eye- and ear-witness Lichtenberg (once more poking fun at the itinerant lecturer), Berschitz did indeed refer to 'Professor Francolini' in his show.[84] In the second half of the eighteenth century, there

79 Lichtenberg to Schernhagen, 17 October 1782; Joost and Schöne, *Lichtenberg, Briefwechsel*, p. 448.

80 Fürstenfeld Führer, § 326, p. 259, 'Mechanicus – aber nur blosser Mechanicus; denn er ist sogar des Schreibens unkundig gewesen'; Kremsmünster, § 188, p. 344, 'virum, qui tamen nec literis studuit'; Haubelt, 'Vznik Prirodovedeckych', p. 79, 'vir sine ullo studio physices'.

81 Another example would be Jakob von Bianchy; Hochadel, 'Business'.

82 The classic work is Smith, *The Body*; specifically for the eighteenth century: Bertucci, *Artisanal*.

83 'Des berühmten Herrn Franklins Theorie von der Elektrizität wird er durch eine besondere Maschine, und Experimente deutlich vor Augen stellen'.

84 Lichtenberg to Reimarus, 2 May 1782; Joost and Schöne, *Lichtenberg, Briefwechsel*, pp. 317–18.

were roughly speaking two competing theories attempting to provide a framework for the multitude of electrical phenomena. The first, as proposed by 'the famous Mr Franklin' and others, proclaimed that there was only one kind of electricity; the second theory postulated two different kinds of electricity. In a slightly vicious way, Lichtenberg turned this against Berschitz, too. The fact that Berschitz was able to understand Benjamin Franklin's theory well and to produce complex experiments proves the 'noble simplicity' of the unitarian theory.[85] Lichtenberg himself tended towards a dualistic view of electricity.

In trying to combine the few bits and pieces the sources offer us on Berschitz's horizon of knowledge, several examples emerge. He correctly claimed that the thunder-house was 'recently invented in England'.[86] Hence he was well acquainted with the Enlightenment discourse on lightning rods and also 'up to date' with the possible applications of the new gas chemistry, as shown above.

Berschitz's sources of information are nearly impossible to pin down. It seems therefore helpful to stress that his professional career was characterised by nearly constant movement, travelling from town to town and encountering different microcosms of learning all the time. The fact that Berschitz actively searched for contact with scholars such as Brander or Lichtenberg (including writing letters to the Göttingen physicist) suggests that he received relevant scientific information through personal contacts rather than through reading journals or books.

The Precarious Status of an Itinerant Lecturer

All of Berschitz's announcements and handbills start by claiming that he is in possession of references from princely courts, magistrates, or other political authorities. This was a standard trope deployed by all itinerant lecturers. They were in need of legitimising themselves. In their pleas for permits, they would submit these references to the magistrates – and hastily add that they would need them back. That is the main reason why hardly any of these references have been preserved.

85 Lichtenberg to Johann Daniel Ramberg, 8 April 1782; Joost and Schöne, *Lichtenberg, Briefwechsel*, p. 303, 'edle Einfalt'.

86 Handbill, see n. 12. One of the first models of this kind 'was marketed by the London instrument maker Edward Nairne and publicised by the itinerant lecturer James Ferguson in 1770'. Hackmann, 'Lightning Rod', p. 216.

Berschitz called himself a 'Mechanikus'. In his advertisements, he often added a title that commanded respect, for instance 'Kaiserl. und Churpfälzischer Hofmechanikus', asserting that he had received an endorsement from a princely court. Apart from drawing on 'traditional' sources of reputation from political authorities Berschitz also felt the need for some kind of academic legitimation. In its letter of recommendation, the magistrate of the Swiss city of Solothurn (the only recommendation preserved in the case of Berschitz) called him 'a studied physicist', presumably because he pretended to be one.[87] While in the Austrian Netherlands, in the city of Leuven in 1776, Berschitz even claimed to be a member of several academies of science; in his advertisement in Mechelen (December 1783), he called himself a physicist.[88] Maybe the fact that he had moved outside the German-speaking area made him more audacious in his advertisement in Dutch, as he never alleged this kind of membership in his announcements in German.

Berschitz also asked the scholars he met to write him reference letters – and was occasionally quite insisting. While in Augsburg in 1773 he contacted the celebrated instrument maker Georg Friedrich Brander (1713–1783), a founding member of the Bavarian Academy of Sciences in the late 1750s. Brander reported in a letter that Berschitz wanted a recommendation 'par force' – but he told him that he would not issue any out of principle. He informed him that he had even rebuffed one of his relatives who had asked for a letter of recommendation to the Polish King.[89]

Berschitz had more success when he contacted Lichtenberg in Göttingen, probably in March 1782. Lichtenberg must have heard about Berschitz's explosive sojourn at Kassel a few months earlier and was certainly curious about him. He recommended Berschitz's performance in his lecture to his students. Berschitz's show at the *Kaufhaus* eventually attracted seventy-four spectators. Lichtenberg gave Berschitz a 'nice electrical tube' and 'other trifles' as a present and considered him a skilled experimenter. Eventually, Lichtenberg wrote him a letter of recommendation, saying, 'that even connoisseurs would see some of his experiments with pleasure'. Yet once Berschitz had left Göttingen their relationship turned sour, at least in Lichtenberg's view. He complained that Berschitz had made inappropriate use of his recommendation in an article published in a newspaper in Altona

87 Copeyenbuch, 24 March 1775, pp. 62–63, Stadtarchiv Solothurn, 'einen studierten Phisiker'.
88 Leuven: 'zynde Lid van verscheyde Academien der Wetenschappen'. See the three announcements in *Wekelijks nieuws uit Loven, mede beschrijvinge diër stad*, 10 November 1776, p. 304; 17 November 1776, p. 319; 24 November 1776, pp. 335–36; Mechelen: *Wekelyks bericht voor de provincie van Mechelen*, 14 December 1783, p. 658, 'natuurkundige'.
89 Brander to Prosper Goldhofer, 3 September 1773; Winkler, *Georg Friedrich*, p. 245.

(Hamburg). Presumably, Berschitz boasted too much, trying to attract a large public to his upcoming performances.[90]

In 1792, performing in Kremsmünster, Berschitz claimed to be well acquainted with the Premonstratensian's canon and natural philosopher Prokop Diviš (1698–1765) and to know his electrical experiments.[91] Diviš was indeed well known at the time, certainly to Doberschiz, and has on occasion been considered a co-inventor of the lightning rod. He had been in Vienna before or around 1750 where he had carried out experiments, again in the presence of the Imperial family, with Franz.[92] That might explain why Berschitz knew of Diviš but it is unlikely that he knew him personally, given the fact that Diviš died in 1765. It might well be that Berschitz wanted to impress the monks at Kremsmünster, as well as Lichtenberg, with his claim. All these episodes clearly show that Berschitz understood the significance of personal reputation in the Republic of Letters. Mentioning the names of Franz, Diviš, Brander, and Lichtenberg was supposed to elevate his status.

Berschitz's attempts to enter the enlightened world of learning, however, were repeatedly rebuffed. Some scholars denounced him outright, at least in their private letters. Brander categorised him as a 'Fischiko', a pseudo-physicist.[93] The naturalist Georg Forster (1754–1794), at the time professor at the Collegium Carolinum in Kassel, dubbed Berschitz the 'miserable Viennese wanna-be-professor' who 'degraded' experimental physics 'to a hocuspocus'. Forster explicitly referred to his sojourn in Kassel in 1781/82 where he might have witnessed his shows.[94] Lichtenberg, always very expressive, particularly when it came to slander, called him the 'greatest ignoramus' and an 'exuberant rake' but also an 'Elektrophorus', an allusion to the electrical instrument that had become popular at the time and literally means 'bearer of electricity' – an epithet that applied well to Berschitz.[95] Although the insults varied, the underlying punch was the same: Berschitz was pretentious, deceitful, and had certainly no claim to be part of the learned community of the German Enlightenment. This chimed with the

90 Lichtenberg to Reimarus, 2 May 1782; Joost and Schöne, *Lichtenberg, Briefwechsel*, p. 317. The newspaper article in the *Altonaër politischen Zeitung* could not be identified.

91 Kremsmünster, § 188, p. 344.

92 Pelzel, *Abbildungen*, pp. 174–75.

93 Brander to Prosper Goldhofer, 3 September 1773; Winkler, *Georg Friedrich*, p. 245.

94 Georg Forster to Anonymous, 20 November 1782, Stargardt, *Autographen*, p. 154, 'elende Wienerische mock-professor'; 'zu einem hocuspocus herabwürdigt'.

95 Lichtenberg to Reimarus, 2 May 1782; Joost and Schöne, *Lichtenberg, Briefwechsel*, p. 317, 'unüberschwenglichste Windbeutel'.

accusation of his inappropriate and uneducated way of talking. Brander claimed that Berschitz was so talkative, he would make himself known merely through his incessant chatter.[96] Lichtenberg quipped that 'his explanations, delivered in Austrian peasant lingo' are 'truly entertaining'.[97]

Public performances of natural phenomena touched directly upon issues of social order, as historians of science have shown.[98] Who was entitled to teach natural philosophy and reveal the inner workings of nature? Was there some kind of supervision needed? How could the project of Enlightenment be maintained in the marketplace? Was there a danger of reviving superstition by presenting astounding and mind-baffling phenomena to an ignorant public? The question of social control is particularly pertinent with respect to itinerant lecturers and their mixed reputation, tainted by their economic motivation, lack of proper education and manners, tendency to incorporate little tricks, and often dubious moral stature.

Lichtenberg has been described as a sort of watchdog, a 'scientific policeman', exposing impostors and staging himself as a defender of enlightened values.[99] Issues of expertise and authority were central to eighteenth-century natural philosophy, which was still eager to show that experimentation produced reliable knowledge. The introduction of the lightning rod at the very end of the century illustrates this competition for authority because it pits once more the professors against the itinerant lecturers.

The Lightning Rod: Controversial Knowledge

'I have learnt to resist thunder'. When in 1784 Lichtenberg enumerated the great discoveries of the eighteenth century, the lightning rod was second on his list.[100] By that time its utility was – in principle – undisputed among scholars and political authorities in the German lands. In fact, they forcefully advocated the introduction of lightning rods.[101] This public crusade in the form of publications and the backing of the lightning rod by many worldly authorities around 1783–1784 opened up a significant new market

96 Brander to Prosper Goldhofer, 3 September 1773; Winkler, *Georg Friedrich*, p. 245, 'masive Beredtsamkeit'.

97 Lichtenberg to Reimarus, 2 May 1782; Joost and Schöne, *Lichtenberg, Briefwechsel*, p. 318.

98 Schaffer, 'Natural Philosophy', p. 11; Schaffer, 'Consuming Flame'; Stewart, *The Rise*.

99 Oettermann, 'Erläuterung'; Hochadel, '"Martinus Electrophorus"', p. 161.

100 Lichtenberg, 'Vermischte Gedanken', p. 930; For an overview of the historiography see Hochadel and Heering, 'Introduction'.

101 Hochadel, '"In Nebula"'; Schock, 'Donnerstrahl', p. 92.

for itinerant lecturers, local instrument makers, but also a number of physics professors. Erecting lightning rods seems to have been part of Berschitz's portfolio since at least 1779, when he replaced the rods on the powder magazines in Bern. Yet at that stage it seems to have been rather an isolated commission. This was soon to change.

Berschitz used the publicity of his shows to offer his services as regards the new protective technology to well-off individuals and the municipal authorities. He was particularly active in the late 1780s. The records are quite dense for the period between the summer of 1788 and the summer of 1789. Within twelve months he put up lightning rods in Allgäu and Swabia (largely speaking in the area between Stuttgart and Munich), in Kempten (August 1788, town hall and other buildings), Isny (September 1788, church tower), Ulm (November 1788, tower of the minster), Augsburg (April 1789, proposal rejected), Kirchheim (Reichsgräflich Fuggerische Schloß around May 1789),[102] Obermedlingen (early June 1789, church tower of the monastery), and Nördlingen (late June 1789, house of the mayor Tröltsch among others). Another intense phase of lightning rod installations has been documented, following his performance at the Strahov Monastery in Prague. In 1795/96 Berschitz erected lightning rods in the Bohemian towns of Pisek, Plzeň / Pilsen, and Klatovy/Klattau.[103]

The highly profitable and hence competitive market of lightning rods quickly generated problems of its own. As there were different ways of designing and installing a lightning rod, debates on both a theoretical and a practical level often ensued. The discussions concerned, for instance, the materials used, the shape of the top, the required thickness of the rod, and the question of how the rod should be fixed to the building and connected with the ground. In fact, controversies about the correct lightning rod design were so widespread in the final two decades of the eighteenth century that one might even consider them the norm, rather than an exception.[104] That Berschitz was involved in a number of these often very harsh arguments, both as a critic and as the one being attacked, is not surprising.

In the following, I shall give a brief overview of these confrontations. On several occasions Berschitz claimed that the existing lightning conductors were dangerous and should be replaced by his own: in Bern in 1779, in

102 *Schwäbische Chronik*, Nr. 48, 22 April 1789; *Augsburgische Ordinari Postzeitung*, 24 June 1789, p. 3.

103 Sedláček, *Dejiny Královského*, pp. 29–30.

104 Purfleet Controversy in England: Home, 'Points or Knobs'; debate in German journals: Schock, 'Donnerstrahl', p. 98.

Augsburg in 1789, and in Amberg in 1790.[105] In Bern, he seems to have convinced the city council as he presented plans in written form to explain his own design. Yet in Augsburg and Amberg his attempts to question the safety of the existing lightning rods backfired. In Augsburg, the first lightning rod had been installed by the instrument maker (including electrical machines) Jakob Langenbucher (1738–1791). Langenbucher, a former apprentice of Brander, could present testimonies that convinced the magistrate to approve of his contraption. In Amberg, Berschitz denounced the lightning rod that had been installed by the professor of the local school Gabriel Knogler (1759–1838). Knogler defended his construction by referring to Johann Jakob Hemmer (1733–1790), whose design he had followed and whom he even had corresponded with. Hemmer, a naturalist at the court of Palatine Elector Karl Theodor in Mannheim, was possibly the highest regarded authority with regard to lightning protection in southern Germany at the time.[106] In both cases, in Augsburg and Amberg, the magistrates reprimanded Berschitz for his 'calumnies' against the citizens of their community.

What about Berschitz's own lightning rods? In early 1782, he offered Johann Daniel Ramberg (1732–1820), the secretary of the Kriegskanzlei (War Chancellery) in Hanover, to erect lightning rods. Unfortunately for Berschitz, Ramberg asked Lichtenberg for his expert opinion on the proposal. Lichtenberg alerted Ramberg that Berschitz owed his lightning rods 'to a travelling Italian', insinuating that they might not be reliable yet without giving any specifics on what might be wrong with his design.[107] Instead, Lichtenberg recommended the service of his own instrument maker, Johann Andreas Klindworth (1742–1813) from the University of Göttingen, who would install the lightning rod in a 'masterly way' for half the price.

A few months after Berschitz had installed a lightning rod in November 1788 on the tower of the Minster in Ulm, complaints were voiced questioning the design – and hence the security provided. To make matters worse, some of the rods had even broken off.[108] Eventually, a new lightning

105 Bern: Ratsmanual, vol. 350, 29 November 1779, p. 193, Stadtarchiv Bern; Augsburg: Hochadel, '"Hier haben"', pp. 154–63; Amberg: Pro Memoria Gabriel Knogler, 7 March 1790, Staatsarchiv Amberg, Fürstentum Obere Pfalz, Regierung Amtsbücher und Akten 1053.

106 Karl Theodor was known for promoting the sciences and ordered Hemmer already in the mid-1770s to install lightning rods in his territories around Mannheim; Voss, 'Carl Theodor', p. 19; Cappel, *Das Wetter*, p. 20.

107 Lichtenberg to Johann Daniel Ramberg, 8 April 1782; Joost and Schöne, *Lichtenberg, Briefwechsel*, p. 303, 'die er einem herumreisenden Italiäner zu verdancken hat'. As Maddaluno, 'Unveiling Nature', p. 65, writes: 'In fact, in the context of exhibitions, "Italian" was synonymous with "charlatan", namely, with a person who was able to deceive the audience with fair words'.

108 Stadtarchiv Ulm A [1619], No. 5,7,8. For more details see Hochadel, 'Fußtruppen', pp. 334–35.

rod system was installed in the spring of 1789 under the supervision of Joseph Weber (1753–1831), professor of physics at the University of Dillingen.[109]

In June 1789 in Nördlingen, while canvassing for his lightning rods, Berschitz felt that he needed to defend himself against 'various rumours spread against him'. He probably referred to the criticism he had received in Ulm and the polemic he had caused in Augsburg just a few months earlier. Therefore, he announced a 'small writing' in order to compare 'all sorts of lightning rods' with those of his own design. We cannot even be sure whether this writing was ever published, but Berschitz apparently felt capable of giving a critical overview of current types of lightning rods. Obviously, the idea was to prove the superiority of his design, yet he declared that he would leave 'the judgment to the public'.[110] This appeal to the public as the final arbiter shows once more how well Berschitz was versed in Enlightenment rhetoric.

So there emerges a pattern: Berschitz's credibility was questioned and the magistrates tended to trust their citizens – and the professors with academic credentials: Lichtenberg in the Hanover case, Weber in Ulm, and Hemmer in Amberg. Yet Berschitz was not always on the losing side of the controversy. In 1796 the local authorities questioned the safety of the lightning rod he had erected in the Bohemian town of Pisek. In his defence, Berschitz contacted the responsible office in Prague which, after a series of arguments, overruled the local authorities. Hennet, the person in charge, invoked two reasons: Berschitz's considerable experience (he had also installed lightning rods in a number of other Bohemian towns) but also the fact that Berschitz was 'from Vienna'.[111] This might be understood as a typical gesture of deference reflecting the epistemological/geographical hierarchies of the Habsburg Empire at the time.

Later that year, in October 1796, Berschitz announced that he had installed a lightning rod in Hetzendorf, at the time just outside Vienna's city border. In a newspaper article, he emphasised that he received this commission through the recommendation of a certain 'Herrn v. Mälzer'.[112] It could not be determined who this 'well-known lover of knowledge of physics' was. Yet the phrasing of the advertisement suggests that Berschitz was fully aware of the need to legitimise his practice as a lightning rod expert by acknowledged naturalists.

109 *Schwäbische Chronik*, 5 June 1789. On Weber see Hochadel, '"Physiker, Aufklärer und "Experte"'.

110 *Augsburgische Ordinari Postzeitung*, 24 June 1789, p. 3.

111 Sedláček, *Dejiny Královského*, p. 30.

112 'Nachricht', *Wiener Zeitung*, 30 November 1796, p. 19 (p. 3421).

The argument of this section is not that the professors got the upper hand in these controversies merely because they *a priori* commanded more authority than itinerant lecturers. Rather, the debate about the proper design forced them to enter the public sphere, argue their case in the negotiations with the political authorities, and thus gradually establish their authority as recognised experts in the field of lightning protection. The example of Berschitz shows though, that itinerant lecturers kept trying to hold on to their share of the market, even if this proved to be an uphill struggle.

Conclusion

Martin Berschitz was not a devotee of science in the sense that he practised natural philosophy in his spare time. Demonstrating electrical phenomena to paying audiences was his profession. Yet to portray him merely as a 'scientific salesman' with little else but thoughts of economic gain, or as a showman more interested in crowd-pleasing tricks than in creating new knowledge, would replicate the argument of natural philosophers such as Lichtenberg. This is not to downplay his economic motivation, but rather to the contrary. Enlightenment was also a business, to paraphrase Robert Darnton.[113] This is as true for the mass publication of the *Encyclopédie* as for the material culture of natural philosophy (publications, instruments, and public demonstrations). For Berschitz, Enlightenment was certainly a business opportunity, a way to carve out a living for himself and his family.

Once he had embarked on this path, he dedicated his entire professional career to performing scientific demonstrations, making and selling instruments, and installing lightning rods. Itinerant lecturers such as Berschitz were instrumental in the spread of insights from natural philosophy in a number of ways, both materially and intellectually. In his thirty years performing and lecturing, he reached thousands of people. He addressed a multitude of audiences, spectators in taverns and the marketplace, school children and monks, but also other natural philosophers and amateurs of science such as Franz Ferdinand Wolff. Berschitz provided both entertainment and instruction, undoubtedly raising interest in electricity and other fields of natural philosophy among his publics. He built and sold a considerable number of instruments, and equipped cabinets with electrical machines, Leyden Jars, and a host of related equipment, providing the material bases for teaching physics and pursuing research.

113 Darnton, *The Business*.

Berschitz's portfolio evolved over time, incorporating 'new experiments, never seen before', new services, and products such as electrotherapy and lightning rods. This underscores that the scientific culture of the Enlightenment was also a marketplace, where itinerant lecturers had to offer new products to keep attracting audiences and customers. Berschitz's interest in new developments and discoveries in the field of natural philosophy was conditioned by his quest for possible applications. These additions to his portfolio could prove more or less successful as demonstrated in the case of the new gas chemistry, the automata, the balloon business, or the demonstration of the thunder-house.

This leads to the issue of how Berschitz was connected to the circulation of scientific and technological knowledge in the late Enlightenment. At first glance, he appears to have been a bit of a lone warrior, rather wrangling and fighting with local authorities and other practitioners from the field of natural philosophy, in particular with respect to the lightning rod. Some of his attempts to obtain references turned out to be futile or backfired on him. He claimed to be acquainted with famous naturalists to convey the impression of being well connected and trained and thus to sway his learned audiences.

He clearly kept abreast of new developments within natural philosophy, in particular with respect to new technologies. Due to the scarcity of sources, it is very hard to say through what channel he learnt about 'dephlogisticated air' or the chess automaton. Constant travelling and personal communication (hardly free of friction) with other instrument makers and scholars conditioned his learning in all likelihood much more than reading. For logistical reasons alone, he must have been able to gather sufficient information to keep going from town to town. The fact that at least for some periods he travelled and performed in non-German-speaking areas testifies to his capacity to adapt. He certainly impressed the Landgrave of Hesse-Cassel, posed as court mechanic of Kurpfalz-Bayern, and could convince a number of municipal authorities of his abilities. The large number of monasteries that became his clients also suggests at least some rudimentary connections helped him to pursue his business. Yet clearly, if one may speak of Berschitz's networks at all, they were fragmentary, transient, and prone to dissolve given the numerous controversies he was involved in.

Berschitz oscillated between short-lived celebrity and admiration (his underwater explosions), and an overwhelmingly sceptical if not outright hostile reception from the learned world, alleging that he contributed no new knowledge, doubting his morals, and ridiculing his manners. Yet everybody who watched him perform could not help but admire his dexterity and skills as a demonstrator.

What shines through the very patchy historical record is a forceful but also opportunistic personality, endowed with grit and resilience, well capable of seeking his advantage and defending his interests in a robust way. Berschitz kept his business afloat for decades, despite numerous practical problems and challenges as well as the widespread disparagement of his profession as an itinerant lecturer. Although we might feel closer to the witty and smart Lichtenberg than to the pompous and bragging Berschitz, both represent different but complementary elements of late Enlightenment science.[114]

Bibliography

Baader, B., 'Merkwürdigkeiten des Reichsstiftes Elchingen von 1785–1818', Historischer Verein für Schwaben, Stadtarchiv Augsburg, H 150 1/2, vol. 1., pp. 364–65.

Baumer-Müller, V., 'Die Gebrüder Abt. Gerold II. Meyer von Muri, Abt Bernhard III. Mayer von Rheinau und Ratsherr J.R. Valentin Meyer: Ein Beitrag zur Situation von Benediktinerklöstern in der Schweiz des ausgehenden 18. Jahrhunderts', *Studien und Mitteilungen zur Geschichte des Benediktiner-Ordens und seiner Zweige* 117 (2006), pp. 345–81.

Bensaude-Vincent, B., and C. Blondel, ed., *Science and Spectacle in the European Enlightenment* (Aldershot 2008).

Bertholon, P., *De l'électricité des météores* [...] *météorologie électrique* [...], vols. 1–2 (Paris 1787).

Bertucci, P., 'Sparking Controversy: Jean Antoine Nollet and Medical Electricity South of the Alps', *Nuncius: Annali di Storia della Scienza* 20, no. 1 (2005), pp. 153–87.

Bertucci, P., *Artisanal Enlightenment: Science and the Mechanical Arts in Old Regime France* (New Haven 2017).

Binhack, F., *Geschichte des Cisterzienserstiftes Waldsassen von der Wiederherstellung des Klosters (1661) bis zum Tode des Abtes Alexander (1756): Nach Manuscripten des P. Dionysius Hueber von Franz Binhack* (Regensburg [u.a.] 1888).

Bloemer, J., *Empirie im Mönchsgewand: Naturforschung in süddeutschen Klöstern des 18. Jahrhunderts* (Göttingen 2022).

114 This paper is indebted to a considerable number of archivists, too many to mention here individually. A number of fellow historians provided me with valuable information on Berschitz in the course of the years: Nikolaj Bijleveld, Julia Bloemer, Jana Černá, Arjen Dijkstra, Karsten Gaulke, Siegfried Kett, Peter Konečný, Andrea Linnebach, Stephan Oettermann, Jan Pařez, Heike Richter, Georg Schrott, Holger Steigerwald, Ernst Strouhal, Marie Thébaud-Sorger, and Peter Winkler. My colleagues from the Institució Milà i Fontanals in Barcelona discussed a draft of this paper. Gràcies! This paper would not have been written without the kind insistence of Samuel Gessner.

Bret, P., 'Un bateleur de la science: Le "machiniste-physicien" François Bienvenu et la diffusion de Franklin et Lavoisier', *Annales Historiques de la Révolution Française* 338 (2004), pp. 95–127.
Buchner, E., *Das Neueste von gestern: Kulturgeschichtlich interessante Dokumente aus alten deutschen Zeitungen*, vols. 1–5 (Munich 1911–1913).
Cappel, A., 'Das Wetter und seine Aufklärer. Johann Jakob Hemmer in Mannheim', *Photorin. Jahrbuch der Lichtenberg-Gesellschaft* 10, no. 1 (1986), pp. 14–26.
Coester, A., and E. Gerland, *Beschreibung der Sammlung astronomischer, geodätischer und physikalischer Apparate im Königlichen Museum zu Cassel* (Kassel 1878).
Darnton, R., *The Business of Enlightenment: A Publishing History of the Encyclopédie 1775–1800* (Cambridge 1979).
Delbourgo, J., *A Most Amazing Scene of Wonders: Electricity and Enlightenment in Early America* (Cambridge 2006).
Dross, F., 'Gottes elektrischer Wille? Zum Düsseldorfer "Blitzableiter-Aufruhr" 1782/83', in: J. Engelbrecht and S. Laux, ed., *Landes- und Reichsgeschichte: Festschrift für Hansgeorg Molitor zum 65. Geburtstag* (Bielefeld 2004), pp. 281–302.
Felderer, B., and E. Strouhal, *Kempelen – zwei Maschinen* (Vienna 2004).
Fellöcker, S., *Geschichte der Sternwarte der Benediktiner-Abtei Kremsmünster* (Linz 1864).
Gütle, J.K., *Beschreibung eines mathematisch-physikalischen Maschinen- und Instrumenten-Kabinets: Mit zugehörigen Versuchen, zum Gebrauch für Schulen*, vols. 1–2 (Leipzig and Nuremberg 1790).
Hackmann, W.D., 'The Lightning Rod: A Case Study of Eighteenth-Century Model Experiments', in: Heering et al., *Playing with Fire*, pp. 209–29.
Haubelt, J., 'Vznik prirodovedeckych sbirek u premonstratu v Praze na Strahove', *Dejiny ved a techniky* 13, no. 2 (1980), pp. 77–85.
Heeren, H., *Ueber die Verehrung Gottes im Gewitter: Eine Predigt am 20. Aug. 1783, als in der Ordnung der biblischen Erklärungen Hiob 37, der Text und zu Vorstellungen dieser Art viel Veranlassung war* (Bremen 1783).
Heering, P. et al., ed., *Playing with Fire: Histories of the Lightning Rod* (Philadelphia 2009).
Hellyer, M., *Catholic Physics: Jesuit Natural Philosophy in Early Modern Germany* (Notre Dame 2005).
Hochadel, O., '"Martinus Electrophorus Berschütz": Georg Christoph Lichtenberg und die wissenschaftlichen Schausteller seiner Zeit', *Lichtenberg-Jahrbuch* (1998), pp. 155–75.
Hochadel, O., '"Hier haben die Wetterableiter unter den Augsburger Gelehrten eine kleine Revolution gemacht": Die Debatte um die Einführung der Blitzableiter in Augsburg (1783–1791)', *Zeitschrift des Historischen Vereins für Schwaben* 92 (1999), pp. 139–64.

Hochadel, O., 'Physiker, Aufklärer und "Experte": Joseph Weber an der Universität Dillingen', in: R. Kießling, ed., *Die Universität Dillingen und ihre Nachfolger: Stationen und Aspekte einer Hochschule in Schwaben. Festschrift zum 450jährigen Gründungsjubiläum* (Dillingen/Donau 1999), pp. 729–52.

Hochadel, O., '"My Patient Told Me How to Do It": The Practice of Medical Electricity in the German Enlightenment', in: P. Bertucci and G. Pancaldi, ed., *Electric Bodies: Episodes in the History of Medical Electricity* (Bologna 2001), pp. 69–90.

Hochadel, O., *Öffentliche Wissenschaft: Elektrizität in der deutschen Aufklärung* (Göttingen 2003).

Hochadel, O., 'Zauberhafte Aufklärung: Etienne-Gaspard Robertson zwischen Schaustellerei und Wissenschaft', in: B. Felderer and E. Strouhal, ed., *Rare Künste: Zur Kultur- und Mediengeschichte der Zauberkunst* (Vienna 2006), pp. 3–19.

Hochadel, O., 'The Business of Experimental Physics: Instrument Makers and Itinerant Lectures in the German Enlightenment', *Science & Education* 16 (2007), pp. 525–37.

Hochadel, O., 'Die Fußtruppen der Aufklärung: Umherziehende Elektrisierer im 18. Jahrhundert', in: U.J. Schneider, ed., *Kulturen des Wissens im 18. Jahrhundert* (Berlin 2008), pp. 329–37.

Hochadel, O., 'The Sale of Shocks and Sparks: Itinerant Electricians in the German Enlightenment', in: B. Bensaude-Vincent and C. Blondel, *Science and Spectacle in the European Enlightenment* (London 2008), pp. 89–102.

Hochadel, O., '"In nebula nebulorum": The Foggy Summer of 1783 and the Introduction of Lightning Rods in the German Empire', in: Heering et al., *Playing with Fire*, pp. 45–70.

Hochadel, O., and P. Heering, 'Introduction: Revisiting an Invisible Technology', in: Heering et al., *Playing with Fire*, pp. 1–22.

Hof- und Staats-Schematismus der röm. kais. auch kais.königlich und erzherzoglichen Haupt- und Residenzstadt Wien [...] (Vienna 1800).

Home, R.W., 'Points or Knobs: Lightning Rods and the Basis of Decision Making in Late Eighteenth Century British Science', in: Heering et al., *Playing with Fire*, pp. 98–120.

Joost, U., and A. Schöne, ed., *Georg Christoph Lichtenberg, Briefwechsel* (Munich 1985).

Kett, S., 'Er brachte den Blitzableiter nach Nürnberg: Johann Conrad Gütle – Mechanicus, Schausteller, Elektrisierer und Wunderheiler, Physiker, Chemiker, Lehrer, Buchautor und Versandhändler', *Mitteilungen des Vereins für Geschichte der Stadt Nürnberg* 96 (2009), pp. 177–228.

Kim, M.G., *The Imagined Empire: Balloon Enlightenments in Revolutionary Europe* (Pittsburgh 2016).

Klamt, J.-C., *Sternwarte und Museum im Zeitalter der Aufklärung: Der Mathematische Turm im Zeitalter der Aufklärung (1749–1758)* (Mainz 1999).

Kramer, F., 'Wissenschaft und Streben nach "Wahrer Aufklärung": Ein Beitrag zur Aufklärung im ostschwäbischen Benediktinerkoster Elchingen', *Zeitschrift für bayerische Landesgeschichte* 54 (1991), pp. 269–86.

Lackner, F., *Die Jesuitenprofessoren an der philosophischen Fakultät der Wiener Universität (1712–1773)* [Doctoral Thesis] (University of Vienna 1976).

Lichtenberg, G.C., 'Vermischte Gedanken über die aerostatischen Maschinen', *Göttingisches Magazin der Wissenschaften und Literatur* 3, no. 6 (1784), pp. 930–53.

Linnebach, A., *Das Museum der Aufklärung und sein Publikum: Kunsthaus und Museum Fridericianum in Kassel im Kontext des historischen Besucherbuches (1769–1796)* (Kassel 2014).

Lynn, M.R., *Popular Science and Public Opinion in Eighteenth-century France* (Manchester 2006).

Lynn, M.R., *The Sublime Invention: Ballooning in Europe, 1783–1820* (London 2010).

Maddaluno, L., 'Unveiling Nature: Wonder and Deception in Eighteenth-Century London Shows and Exhibitions', *Nuncius* 27 (2012), pp. 56–80.

Mergl, L., 'Restaurovaná válcová třecí elektrika ve Strahovském klášteře', *Dějiny věd a techniky* 39, no. 1 (2006), pp. 25–29.

Millburn, J.R., *Benjamin Martin, Author, Instrument-Maker, and "Country-Showman"* (Leiden 1976).

Millburn, J.R., 'The London Evening Courses of Benjamin Martin and James Ferguson: Eighteenth Century Lectures on Experimental Philosophy', *Annals of Science* 40 (1983), pp. 437–55.

Millburn, J.R., 'James Ferguson's Lecture Tours of the English Midlands in 1771', *Annales of Science* 42 (1985), pp. 397–415.

Millburn, J.R., and H.C. King, *Wheelwright of the Heavens: The Life & Work of James Ferguson* (London 1988).

Müller, 'Electricität', in: H.M.G. Köster and J.F. Roos, ed., *Deutsche Encyclopädie oder Allgemeines Real-Wörterbuch aller Künste und Wissenschaften* (Frankfurt a.M. 1783), pp. 213–25.

Oettermann, S., 'Erläuterung zum Exponat: Georg Christoph Lichtenberg: Avertissement gegen Jakob Philadelphia', in: U. Joost, ed., *Georg Christoph Lichtenberg 1742–1799: Wagnis der Aufklärung* (Munich/Vienna 1992), pp. 314–15.

Pařez, J., 'Strahov Premonstratensians and Natural Sciences in the Strahov Library', in: M. Knedlik and G. Schrott, ed., *Res naturae. Die Oberpfälzer Klöster und die Gaben der Schöpfung* (Kallmünz 2006), pp. 43–56.

Pärr, N., *Maximilian Hell und sein wissenschaftliches Umfeld im Wien des 18. Jahrhunderts* [PhD Thesis] (University of Vienna 2011).

Pauls, E., 'Der Luftschiffer Franz Blanchard zu Aachen im Jahre 1786', *Mitteilungen des Vereins für Kunde der Aachener Vorzeit* 2, no. 4 (1888/89), pp. 53–60.

Pelzel, F.M., *Abbildungen böhmischer und mährischer Gelehrten und Künstler, nebst kurzen Nachrichten von ihren Leben und Werken*, vol. 3 (Prague 1777).

Priestley, J., *The History and Present State of Electricity: With Original Experiments* (London 1767).

Roberts, L., 'Science Becomes Electric: Dutch Interaction with the Electrical Machine during the Eighteenth century', *Isis* 90, no. 4 (1999), pp. 680–714.

Rüger, A., 'Populäre Naturwissenschaft in Nürnberg am Ende des 18. Jahrhunderts: Reisende Experimentatoren, öffentliche Vorlesungen und physikalisches Spielzeug', *Berichte zur Wissenschaftsgeschichte* 5, no. 3–4 (1982), pp. 173–91.

Schaffer, S., 'Natural Philosophy and Public Spectacle in the Eighteenth Century', *History of Science* 21 (1983), pp. 1–43.

Schaffer, S., 'The Consuming Flame: Electrical Showmen and Tory Mystics in the World of Goods', in: J. Brewer and R. Porter, ed., *Consumption and the World of Goods* (London 1993), pp. 488–526.

Schneider, A., *Georg Christoph Lichtenberg: Penseur* [Doctoral Thesis] (Université de Paris 1955).

Schock, F., 'Donnerstrahl und Eisenstangen: Die Debatte über den Blitzableiter in den Journalen der Gelehrtenrepublik', in: M. Füssel and M. Mulsow, ed., *Gelehrtenrepublik* (Hamburg 2015), pp. 67–99.

Schrott, G., '"Blitzfang" und "Electricier-Machinen": Zur klösterlichen Sach- und Wissenskultur in der Zeit der Aufklärung', *Studien und Mitteilungen zur Geschichte des Benediktiner-Ordens und seiner Zweige* 118 (2007), pp. 283–337.

Sedláček, A., *Dejiny královského krajského mesta Písku nad Otavou*, vol. 2 (Pisek 1912).

Smith, P.H., *The Body of the Artisan: Art and Experience in the Scientific Revolution* (Chicago 2004).

Standage, T., *The Turk: The Life and Times of the Famous Eighteenth-Century Chess-Playing Machine* (New York 2003).

Stargardt, J.A., ed., *Autographen aus allen Gebieten: Auktion am 22. und 23. März 1983* (Marburg 1983).

Stewart, L., *The Rise of Public Science: Rhetoric, Technology and Natural Philosophy in Newtonian Britain, 1660–1750* (Cambridge 1992).

Suay Matallana, I., and J.R. Bertomeu Sánchez, 'François Bienvenu y la popularización científica en la Ilustración: Demostraciones experimentales, entretenimiento y públicos de la ciencia', *Enseñanza de las ciencias* 34, no. 2 (2016), pp. 167–84.

Sutton, G.V., *Science for a Polite Society: Gender, Culture, and the Demonstration of Enlightenment* (Boulder 1995).

Tantner, A., 'Der Wiener Barometermacher', *Wiener Zeitung*, 3 January 2013.

Thébaud-Sorger, M., *L'Aérostation au temps des Lumières* (Rennes 2009).
Thébaud-Sorger, M., 'Le spectacle de science', in: S. van Damme, ed., *Histoire des sciences et des savoirs*, vol. 1: *De la renaissance aux lumières* (Paris 2015), pp. 133–53.
Voss, J., 'Carl Theodor als Förderer der Wissenschaften', in: J. Bahns, ed., *Carl Theodor und Elisabeth Auguste: Höfische Kunst und Kultur in der Kurpfalz* (Heidelberg 1979), pp. 15–19.
Winkler, P., *Georg Friedrich Branders Briefe an das Kloster Polling 1754–1780* (Augsburg 2017).
Wolff, F.F., 'Etwas über Blitzableiter', *Annalen der Physik* 8 (1801), pp. 69–83.
Zanetti, F., 'Curing with Machines: Medical Electricity in Eighteenth-Century Paris', *Technology and Culture* 54, no. 3 (2013), pp. 503–30.

About the Author

Oliver Hochadel is a historian of science and a tenured researcher at the Institució Milà i Fontanals (CSIC, Barcelona).

5. The 'Paysans' Christian Gärtner and Johann Georg Palitzsch: Astronomical Pursuits of the 'Common Man' and the Public Image of Science

Sibylle Gluch

Abstract: The history of astronomy has seen several protagonists of simple social origins, who distinguished themselves by uncommon astronomical knowledge, mostly gained through self-study. This chapter focuses on two Saxon commoners: the yarn merchant Christian Gärtner (1705–1782) and the peasant Johann Georg Palitzsch (1723–1788). It presents an understudied aspect of the so-called popular Enlightenment ('Volksaufklärung'): while its advocates supported the idea of the learned common man, they also defined what 'learned' meant. In this way, their concern for the erudition of the lower social classes combined with a concern for hierarchy and the maintenance of social control.

Keywords: astronomy, Enlightenment, knowledge, hierarchy, society

As Alice N. Walters remarked, astronomy 'is an unusually public science; everyone who can see, can see the Moon, the stars, the planets, the Milky Way, and unusual objects like comets'.[1] On that assumption, it does not seem surprising that through the centuries, commoners or members of the rural population should have looked up at the skies and marvelled at the stars, even if only a few of these stargazers are known by name today. Taking up the case of two Saxon villagers, the yarn merchant Christian Gärtner (1705–1782) and the farmer Johann Georg Palitzsch (1723–1788), this chapter

1 Walters, 'Ephemeral Events', p. 1.

Bijleveld, Nikolaj, Arjen Dijkstra, and Samuel Gessner, eds. *Cultivating and Communicating Natural and Technical Knowledge around 1800: Devotees of Science.*
Taylor & Francis Group, 2025
DOI 10.5117/9789463728577_CH05

examines the mechanisms that propelled such protagonists to public interest and fame, and investigates their perception by the academically trained community, professional and amateur astronomers alike.[2]

Gärtner and Palitzsch provide an especially interesting case insofar as they lived near each other in the rural surroundings of Dresden and actually knew each other. It is said that the older Gärtner introduced the younger Palitzsch to telescopic observations, but rather than joining forces, the two became competitors in astronomical matters. The most intriguing aspect, however, is the contrasting image built up already during their lifetime: while Palitzsch served as a model of true erudition, the perfect embodiment of the enlightened ideal of the 'learned peasant', Gärtner was used to illustrate the opposite, the 'ignorant peasant', talking of things he did not understand.[3]

Rural Biographies

Unfortunately, we do not have many biographical documents on Gärtner and Palitzsch. The bulk of information derives from a biased and not entirely trustworthy compilation titled *Schattenrisse edler Teutschen* (Silhouettes of noble Germans) published in 1784.[4] A contemporary review characterised the anonymous author as a 'young, highly effervescent enthusiast, who, unconcerned about light and shadow, only sees everything in the dazzling colour with which the magic lantern of his youthful imagination beguiles him'.[5] The descriptions in the *Schattenrisse*, which rely on stereotypes rather than life images, serve to depict the good 'Biedermann' in order to inspire emulation and 'thus foster wisdom, love of humanity and the true gentleness of the heart'.[6] Volume three of the *Schattenrisse*, which contains

2 Gärtner and Palitzsch were neither the first nor the only 'peasants' – as they were referred to by the scientific community – to take an interest in astronomy. See: Herbst, 'Bäuerliche Autodidakten', pp. 175–90.

3 For the so-called popular Enlightenment (*Volksaufklärung*) and its striving to educate the rural population, see: Böning, 'Die Entdeckung des gemeinen Mannes', pp. 117–64; Böning, 'Popularaufklärung – Volksaufklärung', pp. 563–81; Böning, 'Die Entdeckung des "Volkes"', pp. 253–82.

4 [Lose], *Schattenrisse*, Gärtner: vol. 3, pp. 119–57, Palitzsch: vol. 3, pp. 193–240.

5 Anon., in: *Allgemeine deutsche Bibliothek*, vol. 56, no. 2 (1784), pp. 528–29: 'Es ist ein junger hochaufbrausender Enthusiast, der unbekümmert um Schatten und Licht, nur alles in der blendenden Farbe sieht, die ihm die Zauberlaterne seiner jugendlichen Einbildungskraft vorspiegelt'. The review refers to the first volume, while Gärtner's and Palitzsch's biographies belong to the third. However, the style in all volumes is the same.

6 So the anonymous reviewer of the second volume of the *Schattenrisse* in: *Allgemeine deutsche Bibliothek*, vol. 64, no. 2 (1785), p. 295: 'Der Autor hat keinen geringern Endzweck, als: ihm

the biographies of Gärtner and Palitzsch, received a much more favourable review than its two predecessors. Despite its diagnosed 'physiognomic knick-knack', 'emblazoned style', and 'overloaded manner of praise', the reviewer actually liked the book for its portrayals of men of humble origin and recognised it as an 'active rebuttal' of those who wished to exclude the poor from higher education.[7] Both author and reviewer reveal themselves as advocates of a general Enlightenment seeking to involve the simple man as, for instance, the yarn merchant Gärtner, who became known 'because by his own inner strength he broke the fetters that birth and manner of life had put on his greater spirit, and enlightened himself through untiring diligence and reflection'.[8] The aspiration for a general, popular Enlightenment shapes Gärtner's and Palitzsch's biographies, which, in the *Schattenrisse*, are not so much seen as individuals, but as representatives of a successful implication of Enlightenment ideas.

Both men came from the rural surroundings of the residential city of Dresden; Gärtner from the village of Tolkewitz, Palitzsch from the village of Prohlis. Gärtner's father was a yarn bleacher and merchant, while Palitzsch's father was a farmer who worked his own land. Both received the usual basic education in reading, writing, and probably some calculating in a village school and both took up their fathers' professions. As the *Schattenrisse* convey, Gärtner's and Palitzsch's interest in astronomy developed during childhood: both boys were fascinated by the night sky, watched the stars, and creatively imagined their own constellations. Palitzsch learned about books on astronomy by fellow country folk, like shepherds and hunters, who also taught him some of the established constellations.[9] Gärtner, on the other hand, found a more conventional instructor in his schoolteacher. At this point it seems apt to note that the *Schattenrisse*, as well as all successive biographies underscore the singularity of Gärtner's and Palitzsch's achievements – and probably justly so. However, the rural population cultivated a vernacular astronomical knowledge imparted by word of mouth as well as broadsheets, almanacs, and calendars.

bekannte gute, edle Männer deutschen Vaterlandes andern zur Nachahmung darzustellen, und edle Handlungen ans Licht zu ziehen, um dadurch Weisheit, Menschenliebe und wahre Güte des Herzens zu befördern'.

7 *Allgemeine deutsche Bibliothek*, vol. 77, no. 1 (1787), p. 226.

8 [Lose], *Schattenrisse*, vol. 3, p. 119: 'weil er durch eigne innere Kraft die Fesseln zerbrochen, die Geburt und Lebensart seinem größern Geiste angelegt hatten, und durch unermüdeten Fleiß und Nachdenken sich aufklärte'.

9 [Lose], *Schattenrisse*, vol. 3, p. 201.

Astronomical Book Knowledge for Laymen

Publications such as the widespread 'The Limping Messenger' (*Der hinkende Bote*), also read by Gärtner's family, profoundly contributed to the transmission of a basic knowledge in different sciences and disciplines including astronomy.[10] This level of knowledge was exceeded by textbooks written for interested laymen on a variety of mathematical and astronomical subjects and published in considerable numbers.[11] It is not known which books Gärtner used to expand his knowledge, but they might well have been similar to the ones Palitzsch read. These included the 'Astronomy's Forecourt' (*Vorhof der Stern-Wissenschafft oder Astronomiae*) written by the successful populariser of sciences Christian Pescheck, who taught mathematics at the secondary school (*Gymnasium*) in Zittau and had already published a number of mathematical textbooks.[12] Pescheck explicitly aimed at a 'simple, easy and clear manner of instruction' in order to enable 'those who are far from these studies to understand this noble science' of astronomy.[13] His elementary textbook offers descriptive explanations on the celestial and terrestrial spheres, their main circles, the different coordinate systems, the fixed stars, the movement of the planets and the nature of comets. The book also informs its readers about the different world systems – Ptolemaic, Tychonic, and Copernican – as well as about the division of time and the different calendar types. It closes with twenty-four 'astronomical problems', such as finding the deviation of a compass needle, determining the locations of the Sun's rising and setting, and discovering the longitude and latitude of a star. Appropriate for his target audience, Pescheck abstains from calculations and teaches the handling of instruments such as globes and astronomical tables in order to solve the tasks at hand. Throughout his book he mentions relevant sources and refers to important names in the history of astronomy, in this manner providing his readers with ample material to continue their studies. All in all, Pescheck's book seems a reasonable starting point for a hopeful beginner.

The author of the *Schattenrisse*, however, notes another book as having been particularly influential in Palitzsch's astronomical education: namely

10 [Lose], *Schattenrisse*, vol. 3, p. 126. Sorel, 'Die Verbreitung', pp. 202–27; Herbst, *Astronomie*, pp. 402–5.

11 Baasner, *Lob der Sternkunst*, pp. 202–6.

12 Pescheck, *Vorhof*, no. 1758 in the inventory of Palitzsch's library (see note 19).

13 Pescheck, *Vorhof*, 'Foreword (*Vorrede*)', unnumbered: 'auf eine solche einfältige, leuchte [!] und deutliche Lehr-Art, [...], die vortreffliche Astronomie oder Stern-Wissenschafft zu elaboriren; Damit auch die jenigen, so von denen Studiis weit entfernet sind, diese noble Wissenschafft begreiffen, und proprio marte erlernen könnten'.

Kindermann's *Astronomy*.[14] This reference is interesting insofar as the author, Eberhard Christian Kindermann (ca. 1715–after 1750) lacked formal education.[15] Kindermann was derided by the established scholarly community, which detested his speculative manner of writing and ridiculed his mistakes, such as his assertion that Saturn's rings were composed of balls.[16] However, whereas Kindermann's publications appalled erudite reviewers, they nonetheless appealed to a considerable number of readers.[17] Hence, his imaginative and speculative prose, which sometimes touches on the mystical, might just have been what laypeople needed, who were accustomed to reading mainly religious texts or so-called 'moral journals'.[18] For Palitzsch, Kindermann provided an entry to astronomy, but the inventory of his library illustrates that his intellectual (and financial) investment exceeded that of many other people. In the course of his life, he assembled a formidable library comprising 3,518 titles and containing many of astronomy's classics, such as Copernicus's *De revolutionibus*, Hevelius's *Historia cometae*, Kepler's *Epitome astronomiae Copernicanae*, and Huygen's *Cosmotheoros*. In addition, the inventory of his estate lists a considerable number of ephemerides, atlases, and astronomical tables, among them Bayer's *Uranometria*, Flamsteed's *Atlas Coelestis*, Kepler's *Tabulae Rudolphinae*, and de la Hire's *Tabulae astronomicae*, the latter in a German translation.[19] It is impossible to know which of these books Palitzsch had actually read, but even so they set him apart from his counterpart Gärtner, who never assembled a collection like this in the first place, and, furthermore, fell into poverty after the outbreak of the Seven Years War. Palitzsch, on the other hand, with a considerable sum of disposable money to spend *ad libitum* invested in his social distinction. His library, as well as its accompanying collections of natural specimens and scientific instruments, not only proved the owner's

14 [Lose], *Schattenrisse*, vol. 3, p. 202. The work consists of two parts: Kindermann, *Vollständige Astronomie* and Kindermann, *Collegium Astronomicum*, nos. 493 and 598 in the inventory of Palitzsch's library (see note 19).

15 We have little information on Kindermann. See: Mulsow, *Freigeister*, pp. 100–103; Gruber, 'Die Kometenerscheinung', pp. 157–69. I thank Doris Gruber for sending me her article. Gruber, *Frühneuzeitlicher Wissenswandel*.

16 Cf. the review of Kindermann's *Vollständige Astronomie*, published in 1744, in the influential *Göttingische Zeitung von gelehrten Sachen* (1744), pp. 894–96; Ludewig, 'Kurtze Berechnung', p. 136. For Kindermann's reception in scholarly journals, see: Gruber, 'Die Kometenerscheinung von 1743/44', pp. 160–68.

17 [Lose], *Schattenrisse*, vol. 3, p. 202. For Kindermann's *Astronomie* being widely read, see: Baasner, *Lob der Sternkunst*, pp. 202–6.

18 Cf. [Lose], *Schattenrisse*, vol. 3, pp. 196–99; Siegert, 'Der "gemeine Mann"', pp. 32–51.

19 For the inventory of Palitzsch's estate: SächsStA-D, 10079 Landesregierung, Loc. 11693/48.

interests in the natural sciences, but also conformed to erudite as well as 'polite' conventions.[20] In amassing his collections, Palitzsch set himself apart from the common, 'vulgar' type of person, who consequently was not admitted to view his treasures. As the *Schattenrisse* declared, neither 'ignorance' nor 'gross prejudice' – meaning neither Palitzsch's fellow peasants nor his own servants – were given the opportunity to see his books, instruments, and natural specimens. By this exclusion Palitzsch adopted guidelines that characterised the eighteenth-century practice of natural philosophy, which 'was often explicitly class in character'.[21]

Practical Astronomy in the Countryside

As Gärtner, eighteen years his senior, introduced Palitzsch to observations through the telescope, the latter profited from the older man's experience.[22] Gärtner, on the other hand, had had to find a path for himself.[23] As luck would have it, the Leipzig fairs, which he attended on behalf of his yarn business, not only provided opportunities for practical astronomy such as the testing of instruments, but also offered an extremely vibrant milieu embracing merchants, artisans, university students, and scholars alike. These social groups interacted at various points, which allowed Gärtner to gather experience in different contexts and establish contacts on different levels with one encounter leading to the other. In this manner, Gärtner by and by entered the realm of practical astronomy. Within this process, three acquaintances became of particular importance: the mechanician and lens grinder, Rose (dates and first name unknown), the optician and well-known maker of telescopes, Johann Christian Baumann (1711–1782), and the mathematician Abraham Gotthelf Kästner (1719–1800). Kästner became a primary addressee for Gärtner regarding his observational activities.

Gärtner observed the heavens as often as he could. For this purpose, he built up a small collection of instruments, which contained different refracting telescopes, the lenses of which he mostly ground himself, a self-manufactured pendulum clock that showed hours, minutes, and seconds, as well as some globes and maps. It is not known whether he owned measuring

20 The term 'polite' commonly refers to a genteel society in eighteenth-century England, but seems to fit very well here. Cf. Walters, 'Conversation Pieces', pp. 121–54; Golinski, *British Weather*, pp. 108–10.

21 [Lose], *Schattenrisse*, vol. 3, p. 237; Schaffer, 'Natural Philosophy', p. 10.

22 [Lose], *Schattenrisse*, vol. 3, pp. 202–3.

23 On Gärtner's contacts and activities, see: Gluch, 'Wissenschaft', pp. 37–65.

devices such as a quadrant; but around 1740 he erected a small, wooden observation tower on the top of his house, which, in about 1756/57, was replaced by a larger wooden one.[24] The *Schattenrisse* mention that Gärtner kept observation journals, which, unfortunately, do not seem to have survived. In any case, Gärtner developed into an avid observer, who also wished to share his sightings. To this end, he devised a strategy that, at least for some time, was very successful and brought him international attention.

Palitzsch, according to the *Schattenrisse*, first went to visit Gärtner on the occasion of the lunar eclipse, which took place in April 1744.[25] A mutual acquaintance, a young scholar by the name of Menert, is said to have taken him there.[26] Just like Gärtner before, Palitzsch was captivated by the telescopic sight of the heavens and consequently started to procure his own instruments. In the beginning, he probably used Gärtner as an intermediary or even bought his first telescopes from him.[27] Palitzsch also exploited auctions and, by taking the already well-connected Gärtner as a helpful starting point, over time established a network of contacts that helped him obtain rarer instruments. Palitzsch proved as much the avid collector as he did with his library. The estate inventory, drawn up shortly after his death, lists 116 astronomical instruments. In addition, 161 optical items are listed, some of them usable for astronomical purposes, and a wide variety of mathematical and physical instruments including an electrostatic machine as well as a large collection of minerals and natural specimens. Here, again the question of money arises, but equally the question of quality.

Palitzsch's scientific apparatus was rated by the inspector of the Electoral Cabinet for Mathematical and Physical Instruments, Johann Gottfried Köhler (1745–1801), an astronomer himself. Among the astronomical instruments, Köhler assigned the highest value, nine thalers, to a 28-inch achromatic

24 Cf. Gluch, 'Wissenschaft', p. 43. The *Schattenrisse* (vol. 3, p. 132) report that the replacement of the small wooden tower was ordered by the Saxon Elector Friedrich August III.

25 The *Schattenrisse* (vol. 3, p. 203) actually states that Palitzsch was in his twenty-second year, and thus infer that the eclipse had taken place in 1745. However, no lunar eclipse was visible in Saxony in 1745, hence, it must have been the one of 1744 – when Palitzsch had indeed not yet completed his twenty-second year.

26 Possibly Georg Adolf Mehner (1723–1784), who was then studying at Leipzig University. He was the son of Johann Gottfried Mehner (1690–1762), who from 1713 onwards was pastor in Leuben, part of the parish of Tolkewitz village, from where Gärtner came. In 1727 the elder Mehner became pastor in Leubnitz-Neuostra, where Prohlis, Palitzsch's home village, was parished. For Mehner, see: *Pfarrerbuch Sachsen*.

27 The inventory of Palitzsch's estate mentions twenty objective lenses by Gärtner: SächsSta-D, 10079 Landesregierung, Loc. 11693/48: 'Astronomische Instrumente', no. 163.

perspective with elastic tube by Ramsden (no. 267).[28] With each valued five thalers, the eight-day pendulum clock, showing hours, minutes, and seconds and a 'Dutch or Galilean perspective, in a gold-lined "Scheikert" (?), and brass setting' (no. 285) came second, followed by a '20-foot astronomical tube' in wooden barrels (no. 94) and a 'perspective with 4 draw tubes and a 6-foot objective lens' (no. 266), each at four thalers. The next valuable items, at three thalers each, were a celestial globe of 28-inch diameter (no. 36), probably by Willem Janszoon Blaeu (1571–1638), a so-called parallactic machine made of wood (no. 81), and two telescopes, one 8 feet made by Rudolph (no. 98), the other 6 feet and with an achromatic objective lens, albeit without an ocular (no. 99). To be sure, contemporary prices for telescopes, clocks, and other instruments varied widely. However, unless Köhler seriously underrated the collection, Palitzsch's exemplars have to be placed at the lower end of the price range.[29] In general, the majority of objects, in fact more than 70 per cent, were assigned minor sums: smaller globes, different types of orreries, small quadrants, some of them made of brass and one of them apparently made by Giovanni Antonio Magini (1555–1617), a micrometre according to the design of Gottfried Kirch (1639–1710), and so on were valued well below one thaler. All in all, these instruments befitted the collector rather than the wholehearted astronomer – a man, interested more in quantity than in quality. In line with the impression of a multi-interested dilettante, Palitzsch never erected an observatory, but observed the sky from an open field.

Sadly, the observation journals which Gärtner and Palitzsch are said to have kept have been lost. An attempted reconstruction of their astronomical activities thus rests on hints in the *Schattenrisse*, a few documents such as letters, and, more remarkably, on reports in newspapers and journals. For Gärtner as well as Palitzsch informed the public about their observations. Both men concentrated on the observation of astronomical events such as eclipses of the Sun and Moon, occultations of the planets and the fixed stars, the movement of Jupiter's satellites and other noticeable phenomena. In this manner, Gärtner watched moon dogs or mock moons, a halo phenomenon, in 1750, the occultation of Jupiter by the Moon in 1751, and the one of Venus by the Moon in 1752. He habitually made drawings of these occurrences and

28 The numbers in brackets refer to the numbers of the respective inventory of Palitzsch's estate.

29 In 1786, the prices of Ramsden's telescopes ranged from 24 shillings for a 'one foot one in a mahogany tube' to 10 guineas for a '30 inch focus with 3 different magnifying powers, on a stand' (McConnell, *Jesse Ramsden*, p. 172). This roughly compares to 6 to 56 thalers. Exchange rates taken from: Günther, *Die Casanova Tour*.

regarding the occultations, also timed ingress and egress.[30] As encouraged by the contemporary scientific discourse, the appearance of comets was carefully recorded and Gärtner, as later Palitzsch, developed the habit of watching out for them.[31] The former succeeded in detecting comets in 1748, 1757, and in the summer of 1758. In December the same year, however, Palitzsch outpaced him – as well as the whole scientific community – in first spotting Halley's comet. Both Palitzsch and Gärtner reported on the comets of 1760 and 1769, and both watched the transit of Venus through the Sun, but only Palitzsch in 1761 seems to have reported on this event; he timed the stages of the passage with a sundial.[32] Palitzsch also wrote about the solar eclipse in 1764[33] and was interested in variable stars such as Algol (β Persei). It is possible that Palitzsch's curiosity in Algol antedated reports from England about the findings of John Goodricke (1764–1786) and Edward Pigott (1753–1825), which reached Dresden by letter from the Saxon ambassador extraordinary to the English court, Hans Moritz von Brühl (1736–1809) in May 1783.[34] Be that as it may, Brühl purposely encouraged further observations and facilitated the reading of a summary account of Palitzsch's findings to the Royal Society, which was subsequently published in the *Philosophical Transactions*.[35] This was a remarkable compliment to Palitzsch's commitment to astronomy – and entirely facilitated by expedient contacts.

Gärtner, on the other hand, enthusiastically engaged in one specific aspect of astronomical activities, namely the production of optical instruments and orreries. While the *Schattenrisse* refer in passing to Palitzsch as making some of his devices himself,[36] Gärtner actually attempted to earn a living from it. After his first encounters at the Leipzig fairs, he deliberately sought to learn how to grind lenses. Next to his contacts with mechanicians and

30 Only three of these drawings seem to have survived: one in a letter, now kept at the Observatoire de Paris (see note 83), another, made for the Saxon Electress, in the Dresden archives (SächsStA-D, 10026 Geheimes Kabinett, Loc. 565/2, fol. 9). The third was reproduced as an engraving – probably substantially more adorned than the original drawing – in the Leipzig-based journal *Das Neueste aus der anmuthigen Gelehrsamkeit* 9 (1757), title illustration.

31 Cf. Waff, 'The First International Halley Watch', pp. 373–411.

32 Palitzsch, 'Von der Venus', col. 525–28.

33 *Dresdner Gelehrte Anzeigen* (1764), col. 169–74.

34 Cf. *Dresdner Gelehrte Anzeigen* (1759), col. 19 as well two letters from Brühl, dated 16 and 30 May 1783, published in the Dresden journal *Für Aeltere Litteratur und Neuere Lectüre* (vol. 2 (1783), pp. 262–64).

35 *Philosophical Transactions* 74 (1784), pp. 4–5. I would like to thank Virginia Mills, archivist at the Royal Society, for her help in locating the relevant documents.

36 [Lose], *Schattenrisse*, vol. 3, p. 206, without specifying the instruments.

opticians in Leipzig, Gärtner found help in Dresden in the person of Johann Gottlob Rudolph (1721–1776).

From 1757 onwards, after having settled in Dresden in 1753, Rudolph was employed at the Saxon court, first as a mechanician, later as the inspector of the Electoral Cabinet of Mathematical and Physical Instruments.[37] It is not known how Gärtner's contact with Rudolph came about. However, Rudolph seems to have provided ample advice in optical matters and, moreover, took in Gärtner's son, Johann Gottlieb (1737–1758) as an apprentice. In March 1755, the latter went to study at Leipzig University, so Gärtner and Rudolph's meeting might have taken place in the early 1750s, perhaps shortly after Rudolph had moved to Dresden. According to the *Schattenrisse*, Gärtner cunningly intended to improve his own knowledge in this way, as Rudolph had not yet disclosed all his secrets.[38] In any case, Gärtner systematically built up his own workshop. He purchased the necessary tools such as grinding machines and, in 1757, was able to advertise the following range of products in the *Leipziger Zeitungen*:

> good telescopes, large and small, celestial and terrestrial, short perspectives, fine conserving glasses, binoculars of exceptional quality, very bright, also all systemata of the heavens, all made of brass and accurate, likewise a jovilabium, how the satellites move around Jupiter, where they stand at all days and at all hours, and when the eclipses happen, and within which degree and sign each satellite moves; all made of brass and accurate.[39]

For a while, Gärtner seems to have operated his new business successfully. As the *Schattenrisse* declare, he sold his optical goods 'by the thousand far and wide'.[40] If this was exaggerated, his jovilabium indeed received a very favourable review in the 1758 volume of the Leipzig-based journal *Das Neueste aus der anmuthigen Gelehrsamkeit*, edited by the highly reputable

37 For Rudolph, see: Schillinger, 'Johann Gottlob Rudolph', pp. 319–21, and the obituary in: *Wittenbergsches Wochenblatt* 47 (1776), pp. 373–77.

38 [Lose], *Schattenrisse*, vol. 3, p. 131; Gluch, 'Wissenschaft', pp. 46–47.

39 *Leipziger Zeitungen*, 19 September 1757, p. 596: 'gute Tubi, grosse und kleine, coelestes und terrestres, kurtze Perspectivgen, feine Conservir-Brillen, Ferngläser von ausnehmender Güte, sehr helle, auch alle Systemata des Himmels, alles von Meßing und accurat, ingleichen ein Jovilabium, wie sich die Satellites um den Jupiter bewegen, wie sie zu allen Tagen und zu allen Stunden stehen, und wenn die Finsternisse geschehen, und in welchem Grad und Zeichen ieder Satelles läuft; alles von Meßing accurat'.

40 [Lose], *Schattenrisse*, vol. 3, p. 139.

and influential Johann Christoph Gottsched (1700–1766), professor of logic and metaphysics. Gottsched recommended the machine as a 'fine piece of art, which is capable of adorning a library' and invited prospective buyers to contact him.[41] This passage is remarkable, as Gärtner had attracted the attention of one of the most influential figures of the German Enlightenment.[42] Furthermore, this acquaintance did not come about by mere chance. Although the source material on Gärtner and Palitzsch is regretfully limited, it is very clear that neither of them wanted to follow their scientific passions in solitude and reclusion. Both actively sought contact with kindred spirits and, being exceptional in their interests among their rural counterparts, attracted attention in higher circles.

Gazing at the Learned Peasants

Both Gärtner and Palitzsch are said to have received numerous visitors ranging from the curious bourgeois to noblemen and including men as well as women. During the Seven Years' War, Gärtner's and Palitzsch's residences became popular destinations for officers from the Prussian and Austrian armies and they received high-ranking personalities, like Prince Heinrich of Prussia (1726–1802).[43] Further illustrious names include the Prince Maximilian Ludwig Leopold of Brunswick-Wolfenbüttel (1752–1785), who during the 1778–1779 campaign in Bohemia developed a particular liking for Palitzsch.[44] On the occasion of the transit of Venus in 1769, Gärtner reportedly entertained five envoys; at times up to 'six carriages full of gentlemen and ladies' lined up in front of his house.[45] This attention, even if occasionally coupled with a genuine interest in astronomical observations, decidedly carried the character of spectacle.[46] While celestial events provided the background and set the scene, Gärtner and Palitzsch

41 Gottsched, 'Nachricht', p. 231.

42 There is ample research on Gottsched. See, for example: Achermann, *Gottsched*; Ball, *Moralische Küsse*; Döring, *Gottsched.*

43 [Lose], *Schattenrisse*, vol. 3, pp. 136, 218.

44 [Lose], *Schattenrisse*, vol. 3, pp. 218–19. In this case, the assertion of the *Schattenrisse* is backed by a dinner invitation from Prince Leopold for Palitzsch, preserved in: NLA WO, 298 N, Nr. 575.

45 [Lose], *Schattenrisse*, vol. 3, p. 135.

46 There is a large amount of literature dealing with the spectacular in the production and perception of early modern science. However, the particular setting of the erudite 'peasant' being as much or even more spectacular than the science practised does not seem to have received much attention. Cf. Bensaude-Vincent, and Blondel, *Science and Spectacle*; Schaffer, 'Natural Philosophy', pp. 1–43. For further literature, see also: Daum, 'Varieties', p. 329.

became actors entertaining their audience in showing off their knowledge and favourite pastimes. Occasionally, the audience hardly understood what these performances were about, for example, when the ladies did not even know the proper denomination for the observatory stage.[47] The anonymous author of the *Schattenrisse* criticised the negative effects of these time-consuming shows on Gärtner's and Palitzsch's regular breadwinning labour, and mentioned that Palitzsch attempted to regulate his availability for these performances.[48] However, as irritating as these callers might have been, both Gärtner and Palitzsch enjoyed the attention, particularly as the Saxon court also took notice of them.

Gärtner astounded the Electoral household on the occasion of the great solar eclipse of 1748 (partially visible in Dresden). As Joseph Anton Gabaleon von Wackerbarth-Salmour (1685–1761), Lord Chamberlain to the Electoral prince Friedrich Christian (1722–1763), reported to the prime minister Heinrich von Brühl (1700–1763), 'some scholars of astronomy' had invited Friedrich Christian to observe this event under their guidance.[49] The prince did indeed wish to attend; therefore the observation was painstakingly prepared by Joachim Friedrich Meyen (1704–1772), the then inspector of the Electoral Cabinet of Mathematical and Physical Instruments and the mathematician Böhme – probably the senders of the invitation.[50] The event was as much a scientific as a society spectacle with the 'entire royal family, the court ladies, and numerous persons of distinction' watching, while the prince was actually reading the times of the stages of the eclipse. Within this general atmosphere of 'amusement' described by Wackerbarth, Friedrich Christian and his brothers were most surprised at meeting two peasants from the vicinity of Dresden, who 'in practical astronomy can stand up to the greatest mathematicians'.[51] The two peasants he was referring to

47 [Lose], *Schattenrisse*, vol. 3, p. 135: Some ladies called the 'Observatorium' (observatory) an 'Orenfatorium'. Also Palitzsch's biography recounts how odd and ridiculous the requests by some visitors were ([Lose], *Schattenrisse*, vol. 3, pp. 226–27).

48 [Lose], *Schattenrisse*, vol. 3, pp. 135–36, 226.

49 SächStA-D, 10026 Geheimes Kabinett, Loc. 3058/5: Wackerbarth to Brühl, 24 July 1748, fol. 4.

50 SächStA-D, 10026 Geheimes Kabinett, Loc. 3058/5: Wackerbarth to Brühl, 27 July 1748, fol. 2. The mentioned mathematician was probably August Gottlob Böhme (1719–1797). For Meyen see: Schillinger, 'Meyen', pp. 214–15. Also another of Gottsched's journals, the *Neuer Büchersaal* 7, no. 2 (1748), pp. 178–81 reported on the Dresden observations.

51 SächStA-D, 10026 Geheimes Kabinett, Loc. 3058/5: Wackerbarth to Brühl, 27 July 1748. Wackerbarth describes the scene thus (fols. 2–4): 'Que vous dirai-je, Monseigneur, du spectacle et de l'amusement, que l'Eclipse du Soleil nous donnû Jeudi passé. [...] Toute la Famille Roiale, les Dames de la Cour et quantité de Personnes de distinction s'etant rendues à 10. heures du matin au Pavillon du Zwinger [...] que ce qui parût le plus surprenant à SS. AA. RR.^{les}, c'a été

were in all likelihood Christian Gärtner and his older brother, with whom he often observed.[52] From this time onwards, Gärtner supplied the court with notes on astronomical occurrences and on 14 September 1757, he was even allowed to show Friedrich Christian and his wife, princess Maria Antonia, the comet (C/1757 R1), which he had discovered three days earlier. Just as Gärtner proudly reported this Electoral favour to the editor of the *Hamburgisches Magazin*, Abraham Gotthelf Kästner,[53] Johann Georg Palitzsch, equally pleased, recounted his dealings with the court in the *Dresdner Gelehrte Anzeigen* in 1764.[54]

Palitzsch was the first to see the long-awaited comet Halley in December 1758, an incident of such importance that he was allowed to communicate his sightings to 'Our Most Gracious High Sovereignty', Friedrich Christian and was ordered to notify the prince of future astronomical phenomena. He received further favours from the Electoral family, for instance, a visit by Friedrich Christian's successor, Friedrich August III (1750–1827) and his newly-wed spouse Maria Amalie (1752–1828) in August 1769 as well as a visit by the Elector's mother, Maria Antonia (1724–1780) and brother, Prince Karl (1752–1781) in July 1779. These events were duly reported by the Dresden-based paper *Historischer Kern Dresdnischer Merkwürdigkeiten*, which served as a kind of court magazine and entertained its readers with regular updates about the movements and actions of the Electoral family, their high-ranking visitors, and entourage.[55] The *Historischer Kern* notes that Maria Antonia and Karl 'entertained themselves for several hours with physical amusements at the home of Palitzsch, a countryman well-known in the sciences', making the character of spectacle that these visits carried clearly discernible.[56]

It is important to heed that this attention was not fed by love for the sciences alone, but equally by curiosity and as such bestowed freely upon

d'avoir connû deux paisans des environs de Dresde, qui en fait d'Astronomie pratique peuvent tenir tête avec les plus grands Mathematiciens'.

52 Cf. Gluch, 'Wissenschaft', pp. 53–54.

53 *Hamburgisches Magazin* 20, no. 3 (1758), p. 348.

54 *Dresdner Gelehrte Anzeigen* 18 (1764), col. 174: Palitzsch describes the solar eclipse of 1 April 1764 'zumahl man die hohe Gnade genossen den im 1758sten Jahre erschienenen merckwürdigen Cometen den 27. Decbr. gemeldeten Jahres, Unser Allergnädigsten hohen Landesherrschafft anzuzeigen, und damahls gnädigsten Befehl erhalten, wenn wieder etwas vorfallen würde, solches gleichfalls zu melden'.

55 *Historischer Kern Dresdnischer Merkwürdigkeiten* 15 (1769), p. 59; 13 (1779), pp. 50–51.

56 *Historischer Kern Dresdnischer Merkwürdigkeiten* 13 (1779), p. 51: 'wo sie [Maria Antonia und Prinz Karl] sich bey dem in Wissenschaften bekannten Landmann Joh. George Palitzschen etliche Stunden mit phisicalischen Ergötzlichkeiten vergnügten und Höchst Dero Wohlgefallen darüber bezeigten'.

unusual subjects. For instance, while Gärtner and his brother were noted by the court in 1748, another commoner, a shoemaker from Franconia, 'even rarer' than the two rural astronomers, aroused the court's astonishment because of his oratorical talent.[57] Furthermore, Gärtner could not have been present at the 1748 observation if he had not known either the inspector Meyen or the mathematician Böhme. Palitzsch, for his part, had his sightings of Halley's comet reported almost instantly in the *Dresdner Gelehrte Anzeigen* by the aforementioned Christian Gotthold Hoffmann. As the latter was a court official and had himself presented astronomical observations to Friedrich Christian,[58] it could be ventured that Hoffmann also facilitated Palitzsch's announcement of his news to the court. It is significant, therefore, that Gärtner and Palitzsch were able to attract the court's interest because of contacts they had already established.

Moving in Circles of Science Enthusiasts

Both Gärtner and Palitzsch took part in scientifically interested circles, which reveal much more about their scientific standing and ambitions than their dealings with the court. The surviving records are patchy and deliver a fragmentary image only. It nonetheless becomes clear that both Gärtner and Palitzsch purposefully connected themselves to like-minded people. Gärtner, in particular, seems to have sought contact with mechanicians and scholars in order to further his change of career from yarn merchant to maker of scientific instruments. As the documented acquaintance with Johann Gottlob Rudolph and the undocumented but inevitable acquaintance with Joachim Friedrich Meyen prove, Gärtner successfully approached relevant individuals in the ambience of the Saxon court, such as the court optician, the court mathematician, the inspector and the senior inspector (*Oberinspector*) of the Electoral Cabinet of Mathematical and Physical Instruments. The latter was Georg Gottlieb Haubold (1714–1772), who in 1771 became professor of physics in Leipzig.

A short narrative in the *Schattenrisse* illustrates the active role that has to be attributed to Gärtner as much as to Palitzsch in establishing suitable contacts, although this story refers to the latter only: Palitzsch, who had

57 SächsStA-D, 10026 Geheimes Kabinett, Loc. 3058/5: Wackerbarth to Brühl, 27 July 1748, fol. 4: 'Après que SS. AA. RR.^{les} eurent questionné long têms ces deux paisans, il se presentâ un troisieme sujet encor plus rare qu'eux'.

58 Cf. *Neue Versuche nützlicher Sammlungen* 1 (1750), p. 630.

heard of Haubold via Gärtner, recognised that this would be a most desirable acquaintance. He therefore sought to obtain 'Haubold's friendship in all possible manner' and 'took the greatest pain to first meet him'. He did not only show Haubold his meteorological observations, but also mollified him with small gifts such as honey and cake.[59] The accumulation of meteorological data, which was part of astronomy's wide spectrum of activities, surely proved an easy entree to the science for people with little formal education. In Palitzsch's case, it opened the way to the much-advanced knowledge of mathematics, physics, and astronomy that Haubold possessed. It also provided access to the respective instruments in the electoral collection. The author of the *Schattenrisse* suggestively claims that, in time, Palitzsch surpassed Haubold in astronomical and physical prowess with the consequence that the latter became jealous of the former's achievements.[60] While this assertion seems more like a punch line serving to highlight the narrative of the extraordinarily learned peasant than a presentation of facts, it nonetheless transpires that Gärtner and Palitzsch were affiliated to a local network of astronomically minded men that contributed to their education including their ability to handle astronomical instruments.

Particularly in the case of Gärtner, this network was not restricted to Dresden, but closely tied to Leipzig University. During his regular visits to the triennial Leipzig fairs, Gärtner met students as well as an illustrious group of academics, among them the aforementioned Kästner and Gottsched, but also Gottfried Heinsius (1709–1769) and Christian Fürchtegott Gellert (1715–1769). Kästner and Heinsius were mathematicians and astronomers, Gellert a poet and philosopher, and Gottsched a pronouncedly pugnacious philosopher, linguist, and literary critic. Heinsius served as professor of mathematics since 1745 after having spent eight years as associate professor of astronomy and adjunct to Joseph-Nicolas Delisle (1688–1768) at the Academy of Sciences in Saint Petersburg.[61] Kästner became associate professor in 1746, in 1756 he followed a call as professor of natural sciences and mathematics to the University of Göttingen.[62] Both Heinsius and

59 [Lose], *Schattenrisse*, vol. 3, p. 205: Palitzsch '[suchte] Hauboldts Freundschaft auf alle Weise zu erlangen [...]. Er gab sich gewaltig viel Mühe, um erst an ihn zu kommen, – brachte ihm auf die freundlichste und demüthigste Weise seine Meteorologischen Beobachtungen – denn mit dem Barometer und Thermometer wußte er schon sehr gut und fleißig umzugehen, – an hohen Festtagen dann wohl eine Scheibe Honig, oder ein Bündel Kuchen und dergleichen, um durch dergleichen Gefälligkeiten erst seine Gewogenheit zu gewinnen'.

60 [Lose], *Schattenrisse*, vol. 3, p. 205.

61 Bruhns, 'Heinsius', p. 656.

62 For Kästner see: Baasner, *Kästner*.

Kästner had also studied in Leipzig. Gottsched went to Leipzig in 1724. In 1730 he was appointed associate professor of poetry, in 1734 full professor of logic and metaphysics.[63] Gellert studied in Leipzig from 1734 to 1739 and again from 1740 to 1744. In 1751 he was appointed associate professor of philosophy; in 1761 he declined an offer of full professor of philosophy.[64] These four were related to each other in many ways. They moved in circles concerned with Enlightenment thinking, which in the 1740s became the dominant discourse at Leipzig University.[65]

Many of its representatives, such as Gottsched and Kästner, combined wide-ranging interests across disciplinary boundaries with a striving to educate the laity. Thus Gottsched's main concern was in literature and language, yet he also acted as 'populariser of science', and translated, for instance, Fontenelle's *Entretiens sur la pluralité des mondes* and Musschenbroek's *Elementa Physica*.[66] In the same manner, his periodicals, especially the *Neuer Büchersaal der schönen Wissenschaften und Künste* (1745–1750) and *Das Neueste aus der anmuthigen Gelehrsamkeit* (1751–1762) continuously disseminated information on astronomical matters such as lunar and solar eclipses, the planets, comets, and so forth.[67] Furthermore, as Heinsius lacked an observatory and could not see much from his own quarters, he used Gottsched's domicile with the latter taking an active part in the observations.[68] Thus, though we do not know exactly when Gärtner came into contact with Leipzig's academic world, we can be certain that he encountered an atmosphere which was generally favourable and open to his efforts.

63 Cf. note 42.

64 Wölfel, 'Gellert', pp. 174–75.

65 Mühlpfordt, 'Zwischen Tradition und Innovation', p. 168.

66 Schatzberg, 'Gottsched', pp. 752–70. Bernard de Fontenelle's highly successful *Entretiens* first appeared in Paris in 1686. Gottsched's translation titled *Gespräche von mehr als einer Welt* was published in Leipzig in 1726. Until 1760, four more editions followed. (Schatzberg, 'Gottsched', p. 753.) The first edition of Petrus van Musschenbroek's *Elementa Physica* was published in Leiden in 1734. Gottsched's translation titled *Hrn. Peters von Muschenbroek M. D. der Weltw. und Mathem. ordentlichen Lehrers zu Leyden, Grundlehren der Naturwissenschaft* was published in Leipzig in 1747. (Schatzberg, 'Gottsched', p. 755.)

67 See the respective entries in the complete registers, each published in the last volume of the periodical in question: *Neuer Büchersaal* 10, no. 6 (1750) and *Das Neueste aus der anmuthigen Gelehrsamkeit* 12 (1762).

68 For example, when observing Halley's comet in the spring of 1759: Letter Heinsius to Euler, 21 April 1759 (n. 63), among other fol. 67a. Gottsched had also observed the solar eclipse of 1748 and entertained a number of 'characterised persons and ladies' with it, because Heinsius as the professional could not be disturbed in his careful observations: *Neuer Büchersaal* 7, no. 2 (1748), pp. 178–83 (181).

While Gärtner apparently delighted in being received by Gellert,[69] it was Gottsched and, in particular, Kästner who fostered his astronomical ambitions, while Heinsius, as we will see later, adopted an ambiguous attitude. From 1747 onwards, Kästner, in cooperation with the physician Johann August Unzer (1727–1799), edited the *Hamburgisches Magazin*, a journal that specialised in natural sciences.[70] In the 'Foreword' (*Vorrede*) of the first volume, the editors declare their intention to report on the activities of the international academies, while at the same time stressing that this did not mean excluding works by their compatriots. On the contrary: 'if anyone possesses or has discovered something worth being noted, supposing it to be even the observation of an intelligent countryman, and if he would communicate this to us under his name, so we will accept it with gratitude and make it known'.[71]

This is exactly what Kästner did with Gärtner's information. From 1750 onwards, readers of the *Hamburgisches Magazin* repeatedly heard of Gärtner's observations of mock moons, occultations of planets by the Moon and comets.[72] Regarding these reports, it seems likely that Gärtner came into contact with Gottsched via Kästner, as Gottsched first mentioned Gärtner in 1757. The appearance of a comet in the autumn of this year prompted some explanations concerning the nature and movements of these celestial bodies in *Das Neueste aus der anmuthigen Gelehrsamkeit*. Explaining the accompanying engraving, which was partly based on a drawing by Gärtner, Gottsched remarks that Gärtner came to Leipzig in order to jointly observe the comet with local scholars.[73] The next year, Gottsched brought a description of Gärtner's jovilabium: this machine showed the movements of Jupiter's four known satellites. By means of a crank, they could be made to circle the planet each in its own particular speed and in accordance with astronomical observations. As mentioned previously, Gottsched advertised the machine and prospective buyers could actually examine it at his house.[74] All in all, in the 1750s Gärtner was on

69 [Lose], *Schattenrisse*, vol. 3, p. 155.

70 *Hamburgisches Magazin*.

71 *Hamburgisches Magazin* 1, no. 1 (1747), 'Vorrede' (n.p.): 'Die eigenen Arbeiten unserer Landesleute wollen wir gar nicht ausschließen; sondern wenn jemand in diesen Wissenschaften etwas Bemerkenswürdiges besitzet oder entdecket hat, gesetzt, daß es auch eine Beobachtung von einem klugen Landmanne wäre und uns solches unter seinem Namen mittheilen will, so wollen wir es mit Dank annehmen und bekannt machen'.

72 *Hamburgisches Magazin* 5, no. 1 (1750), pp. 67–68; 8, no. 1 (1751), p. 62; 8, no. 4 (1752), p. 639; 20, no. 3 (1758), p. 347.

73 *Das Neueste aus der anmuthigen Gelehrsamkeit* 10 (1757), pp. 823–27.

74 Cf. note 41.

friendly terms with a number of Leipzig's professors such as Kästner and Gottsched. During this time Gärtner gained self-confidence, for, from 1757 onwards, he published his sightings in newspapers such as the far-reaching *Leipziger Zeitungen*, something he had failed to do when he had discovered a comet in 1748. In addition, Gärtner reached out far beyond Leipzig and contacted Giovanni Domenico Maraldi (1709–1788) in Paris.

In 1752, Gärtner was already worked at his jovilabium and since he encountered difficulties in programming the movements of the satellites, he wrote to Maraldi asking for help. He hoped that Maraldi could be persuaded to include 'the distance of the moons of Jupiter in their orbit' for every day or at least for every first day of each month in the *Connoissance des temps*.[75] Probably as a token of his astronomical adroitness, he included an account of his observation of the occultation of Venus by the Moon in February the same year.[76] Maraldi was named as calculator on the title page of the *Connoissance des temps*, which explains Gärtner's choice of addressee. The letter underlines Gärtner's resourcefulness and seriousness in astronomical matters. In fact, his sightings of comets and their subsequent publication brought him to the attention of several reputable astronomers. Gärtner, himself, claimed to have been contacted by 'professors from Vienna, Munich, Prague, Berlin and Göttingen' because of his observations.[77] From this list, only Vienna and Göttingen are verifiable: Gärtner communicated with Maximilian Hell in Vienna, while Göttingen was Kästner's home from 1756 onwards.[78] Having said as much, Dirk Klinkenberg in The Hague and Joseph-Nicolas Delisle in Paris, both intensely occupied with comet lore, also grew interested in Gärtner. As international papers such as the *Gazette de Leyde* (*Nouvelles*

75 Only the French translation has been preserved: BOP, Correspondance de Joseph-Nicolas Delisle, B 1/7-47: 'Lettre de Christian Gärtner à Jean Dominique Maraldi, 24 septembre 1752': 'j'aurai une priere a leur faire et princepalement a Mr Maraldi, s'il ne voudrait avoir la bonté de metre la distance des lunes de Jupiter dans leur orbite, tous les ans dans leurs Almanacs si cela ne peut pas etre sur chaque jour ou du moins de la faire metre pour le 1er Jour du mois. […] je prie ces Messieurs de ne me point refuser et [---] faites moi imprimer dans quelle signe et dans quel degré ils sont tous les jours ou tout du moins pour le premier Jour du mois'.

76 This account is now kept separately: BOP, A 4-1 (239): 'Observation de l'Eclipse de Venus par la Lune observée le 11 fevrier 1752 au Soir par Dresde à Dolckewitz (peutetre Bolkhowitz en Silesie) par Christian Gurtner'. However, Gärtner's wording makes it clear that these pages originally accompanied the letter.

77 SächStA-D, 10026 Geheimes Kabinett, Loc. 565/2, fol. 10r: Letter by Christian Gärtner to Elector Friedrich August III, 29 May 1769.

78 Hell refers to a letter by Gärtner: *Leipziger Zeitungen*, 4 February 1760, p. 79. At this point in time, they might have known each other for some years. In 1768, as Hell passed through Dresden on his way to Vardoe, Gärtner offered him a telescope of his making, which Hell, however, did not purchase. See Gluch, 'Wissenschaft', p. 49.

Extraordinaires de Divers Endroits) and the *Gazette de Bruxelles* copied news from the *Leipziger Zeitungen*, international readers such as Klinkenberg and Delisle became aware of Gärtner's observations.[79] Consequently, Gärtner became the subject of frenzied enquiry. Thus Delisle wrote to the astronomer Christian Mayer (1719–1783) in Heidelberg, to Heinsius and to Klinkenberg, while the latter discussed the events with Nicolaas Struyck in Amsterdam before finally approaching Gärtner himself.[80] At this point, however, the limitations of Gärtner's knowledge and activities soon transpired: after his response had reached Klinkenberg, the latter noted with disappointment: 'But it may well be that his observations consist of no determinations of positions, and only in contemplation'.[81] Heinsius, on his part, assured Mayer and Delisle that Gärtner was 'not the great astronomer the world thought'. Indeed, Heinsius even suggested Gärtner might have reported more than he actually saw.[82] While this allegation probably served to veil Heinsius's own shortcoming in observing the comet, Gärtner's fame was in fact short-lived. Before exploring the reasons for his 'downfall', however, I shall look at Palitzsch, who moved in somewhat different circles.

Palitzsch's social contacts seem to have been concentrated in Dresden. Given the fact that he had to work his land and had no business to conduct in Leipzig or elsewhere, this is to have been expected. While both Gärtner and Palitzsch succeeded in forming relationships to scientifically educated men with wide interests, only Palitzsch appears to have fitted the Enlightenment ideas closely enough in order to become accepted as an equal. To identify Palitzsch's contacts, the *Schattenrisse* serve again as the main source, but, as before, have to be read with caution. The text mentions, for instance, a 'doctor and legation councillor (*Legationsrath*)' by the name of Schulze.[83]

79 Gluch, 'Wissenschaft', p. 58. Many thanks to Arjen Dijkstra for his help with the research of Dutch newspapers and to Daniel Simpson for supplying me with photographs of the *Gazette de Bruxelles*.

80 Gluch, 'Wissenschaft', pp. 58–61. I am grateful to Huib Zuidervaart, who gave me his notes on Klinkenberg's correspondence and pointed me to the archive, where it is kept. For Klinkenberg's correspondence with Struyck concerning Gärtner, see Zuidervaart, *Van 'konstgenoten'*, pp. xx. For Delisle see: Schaffer, 'Halley', pp. 254–98.

81 Noord-Hollands Archief, Haarlem, Koninklijk Instituut der Wetenschappen (KNAW), toegang 175, inv.nr. 32: Letter from Klinkenberg to Struyck, 12 December 1758: 'mar het zou wel konnen weezen dat jijne waarneemungen in geen bepaalingen van de plaatsen, en alleen maar in de beschouwinge bestaan'.

82 Archives nationales, Paris, MAR/2JJ/67: Letter from Mayer to Delisle, 20 February 1759. Regarding the defence strategies of natural philosophers see: Schaffer, 'Natural Philosophy', p. 13.

83 The *Schattenrisse* (vol. 3, pp. 207–8) mention that at the beginning of the Seven Years' War Palitzsch sent many of his books to 'his good friend' Schulze in Dresden in order to keep them

The latter, described as a good friend of Palitzsch, might have been the broadly interested physician Christian Friedrich Schulze (1730–1775), who, however, was not a legation councillor, but held the academic title of licentiate (*Licenciat*). Schulze himself spoke of Palitzsch as 'a man experienced in the sciences' in a text about Halley's comet.[84] The *Schattenrisse* further refer to a number of court officials such as Anton Hermann SJ (1694–1768), confessor of the Electress Maria Josepha (1699–1757), and apparently well informed about discoveries and developments in the natural sciences, Johann Christoph Glaser (d. 1773), war councillor (*Kriegsrat*) and professor at the cadet corps, Christian Gotthold Hoffmann, excise taxes councillor (*Akziserat*), Karl Christian Canzler (1733–1786), and Karl Wilhelm Dassdorf (1750–1812), both court librarians.

Through these men, Palitzsch found intellectual stimulation in different fields. Father Hermann and Glaser are described as lovers of the natural sciences.[85] Glaser, about whom not much is known, owned a large library, comprising more than 5,580 volumes, dealing with a wide variety of subjects, including, for instance, military, historical, mathematical, and biological texts.[86] Hoffmann, who had studied law in Leipzig and Wittenberg, was an avid amateur astronomer. He corresponded with Maximilian Hell in Vienna, to whom he repeatedly sent his observations, for instance, of the transit of Venus in 1761 and the occultation of Saturn by the Moon in 1762 as well as extensive meteorological data in 1766.[87] Hell published many of these communications in his *Ephemerides* and elsewhere.[88] Canzler, who was well versed in seven languages (Hebrew, Greek, Latin, English, French, Italian, and Dutch), enthused about the history and literature of the Middle Ages. He had been employed at the library of the powerful prime minister Heinrich

safe. They burnt, however, during the siege and bombardment of the town. The 'licentiate' is mentioned, for example, by Schröter, *Journal*, p. 170 and in the membership lists of the *Leipziger ökonomische Societät*, printed in the *Leipziger Adreß-, Post- und Reise-Kalender*, for example: 1768, p. 73. For Schulze, about whom not much is known, see: Fischer, *Mineralogie*, pp. 248–55 (n. 467).

84 *Dresdnisches Magazin* 1, no. 7 (1760), pp. 433–34. Here, Schulze gives details not printed elsewhere, a fact that supports a personal acquaintance.

85 [Lose], *Schattenrisse*, vol. 3, pp. 206, 210–11. In 1759 Hermann became rector at the Jesuit college in Konstanz. Cf. Duhr, *Geschichte der Jesuiten*, vol. 4, no. 2, p. 326.

86 The inventory of Glaser's library is kept in: SächStA-D, 11332 Kriegsgerichte besonderer Behörden und Formationen bis 1867, no. 921.

87 University Archive Vienna, Astr. NL 2.13: 'Paralactische Vorstellung des Vorübergangs der Venus vor der Sonne [...]'; Astr. NL 2.43: 'Beobachtung der Bedeckung des [Saturn] von dem Monde [...]'; Astr. NL 2.61: 'Baro- und Thermometre Observation'.

88 For example: Hell, *Ephemerides astronomicae anni bisexti 1764 ad meridianum vindobonensem*, pp. 246–48; *anni 1765*, pp. 318, 334; *anni bissexti 1768*, pp. 280–81. On the transit of Venus: Hell, *Observatio*, pp. 82–83. Hell also knew Hoffmann's book *Der gelehrte Bauer* (see note 131).

von Brühl (1700–1763), (in)famous for his disastrous financial politics, and, after Brühl's death, became head librarian of the Electoral library.[89] Dassdorf, who also worked there from 1775 onwards, was a friend of one of the most famous German proponents of the Enlightenment (*Aufklärer*) Gotthold Ephraim Lessing (1729–1781), and himself a prolific writer.[90] He also edited the letters of Johann Joachim Winckelmann (1717–1768), who is nowadays considered to be one of the founders of archaeology and art history.[91] Even though we do not know how these contacts came about, their significance can be estimated as they opened up further contacts and opportunities.

The biographies and professional positions of the men listed above entailed contacts with government officials, intellectuals, and the court. Regrettably, these circles have been researched very little. However, given Dresden's modest size and the similarity of educational biographies of which the University of Leipzig was one focal point, it can be presumed that within this elite group everybody knew everybody, at least through three corners. Thus Hoffmann, Canzler, and Dassdorf had all studied at some point in Leipzig. Naturally, the three of them knew Gellert and Gottsched. Via Hoffmann, Gottsched met another 'learned peasant', the tax collector Johann Ludewig.[92] Also Hans Moritz von Brühl, who had read Palitzsch's observation of Algol to the Royal Society, had attended Leipzig University. Brühl was a close friend of Gellert's, whom he supported financially for many years, and a friend of Canzler's, whom he helped to find his first position as a librarian in Warsaw. Brühl might have been personally acquainted with Palitzsch, but in the 1780s, Canzler acted as intermediary between the two. For example, Brühl informed Canzler that he had unsuccessfully tried to find seeds of the Venus flytrap (*Dionaea muscipula*) for Palitzsch, while Canzler forwarded Palitzsch's observations of Algol to Brühl. In addition, Canzler communicated all these dealings to the interested public in his journal *Für Aeltere Litteratur und Neuere Lectüre*.[93] By this time, however, Palitzsch

89 Nitzschke, 'Canzler'.

90 Nitzschke, 'Dassdorf'.

91 Dassdorf, *Winckelmanns Briefe*.

92 Gottsched mentions this encounter in his *Historische Lobschrift*, pp. 32–33, n. Hoffmann reprinted this passage in the editorial of his *Der gelehrte Bauer*. Although not explicitly said, it seems certain that Hoffmann facilitated this encounter. His acquaintance with Gellert is documented in Gellert's correspondence with Christiane Caroline Lucius, the daughter of Carl Friedrich Lucius, secret secretary and secret cabinet registrar of the *Geheime Kabinett*, the supreme governing body. Cf. Ebert, *Briefwechsel*, pp. 43, 98, 316, 350, 374–78, 380–81.

93 *Für Aeltere Litteratur und Neuere Lectüre* 2 (1783), pp. 259, 262–63; 4 (1783), pp. 239–40; 3 (1784), pp. 244, 246–50.

had become a most respectable individual who enjoyed the friendship of politically highly influential figures such as that of the court and judicial councillor (*Hof- und Justizrat*) Friedrich Wilhelm Ferber (1732–1800).

Ferber was one of the key figures of the Saxon *rétablissement*. This reform movement aimed at rebuilding and restructuring the Saxon economy and administration which were in an acute state of crisis after the end of the Seven Years' War.[94] Supported by the 'young court', the crown prince Friedrich Christian and his wife, Maria Antonia, it was predominantly shaped by Thomas von Fritsch (1700–1775) and Christian Gotthelf von Gutschmid (1721–1798), both of bourgeois origin. Gutschmid, who reportedly dined at Hoffmann's table, formed a whole generation of state officials, among them Ferber.[95] During his studies at Leipzig University, the latter had formed a close friendship to Brühl and Gellert's brother, the fencing master Friedrich Leberecht (1711–1770).[96] After graduating with a doctorate in law in 1755, he was employed as secret chamber secretary (*Geheimer Kammersekretär*) in the supreme governing body (*Geheime Kabinett*), like his father and grandfather before him. From the 1760s onwards, he held several important positions in the financial and economic administration of the Saxon government and, being excellently connected, exerted a tangible influence on the inner affairs of the state. The French ambassador Du Buat-Nançay regarded Ferber a 'great admirer of democracy'.[97] In any case, Ferber's political and economic notions, as well as those of the other proponents of the *rétablissement*, were guided by the ideas of the Enlightenment.[98] We do not know how Ferber became acquainted with Palitzsch. From 1773 to 1775, Palitzsch's friend Dassdorf worked as tutor (*Hofmeister*) in Ferber's house, but Ferber might have met Palitzsch well before that time. Given the efforts that the Saxon *rétablissement* devoted to rebuilding the devastated country including its agriculture, it does not seem surprising that a 'model farmer' such as Palitzsch, who in 1764 designated himself as a 'devotée of agriculture, physics

94 A still nominal account of the Saxon *Rétablissement* is: Schlechte, *Die Staatsreform*. However, Schlechte's view that the *rétablissement* aimed at establishing an enlightened absolutist state has been repeatedly called into question. See, for instance: Matzerath, '"Pflicht"', pp. 157–82 and Keller, 'Saxony', pp. 309–31 (here also further literature).

95 Christiane Caroline Lucius met Gutschmid at Hoffmann's table: Ebert, *Briefwechsel*, p. 316; Schlechte, *Die Staatsreform*, pp. 61–66.

96 This circle of friends would merit further research. Ferber's friendship with Brühl and Friedrich Leberecht Gellert as well as his close acquaintance with Christian Fürchtegott Gellert becomes evident in letters kept at the Dresden state archive: SächStA-D, 12582 Familiennachlass von Carlowitz-Ferber (D), nos. 073, 074, 076.

97 Schlechte, *Die Staatsreform*, p. 63, n. 254.

98 Keller, 'Saxony'.

and astronomy', should attract attention, the more so as Ferber was an early member of the *Leipziger ökonomische Societät*.[99]

The origin of this society was closely connected to the general ideas and efforts of the *rétablissement*. Founded in May 1764 by reason of the 'obligation to work with united efforts for the advancement of the common good', the society aimed at improving the state of everything concerning food production (*Nahrungsstand*).[100] Ferber joined the society on 15 January 1765 as member number 66.[101] Palitzsch was elected associate member in 1770, presumably on account of his farming efforts: in its statute, the society declared that representatives of every class could be elected, 'when they excell'.[102] As an associate member, Palitzsch did not have to pay a membership fee, but, in return, only had a consultative voice. Despite these restrictions, his membership certainly meant an honour as rarely any individuals of his social rank were able to join the society, which was dominated by the 'nobility, state servants and manor owners'.[103] These three groups are united in the person of Friedrich Wilhelm Ferber, who became an official in the late 1750s, was ennobled in 1776, and bought the estates (*Rittergüter*) of Jänkendorf and Kaana in 1782.[104]

In 1779, Ferber asked the artist Christian Gottfried Schulze (1749–1819) whether he would agree to engrave Palitzsch's 'head' after a portrait in Ferber's possession by the acclaimed court painter Anton Graff (1736–1813).[105] At this

99 Palitzsch signed his report on the eclipse of the Sun in 1764 in this manner: *Dresdner gelehrte Anzeigen* 18 (1764), col. 174.

100 'Anzeige von der neu errichteten Leipziger ökonomischen Societät', *Leipziger Intelligenz-Blatt* 24 (1764): *Beylage zum 24sten Stück*, pp. 217–20 (217). See also: Schöne, 'Die Leipziger Ökonomische Sozietät', pp. 53–78.

101 *Leipziger Intelligenz-Blatt* 22 (1765): *Beylage zu dem 22sten Stück* (s.p.).

102 'Anzeige von der neu errichteten Leipziger ökonomischen Societät' (n. 111), p. 219. Palitzsch's election: *Leipziger Intelligenz-Blatt* 50 (1770): *Beylage zum 50sten Stück*: 'Funfzehende Anzeige von der Leipziger ökonomischen Societät', p. 483. Very early on, this election was interpreted as an Electoral favour, although we have no traces of any involvement of this kind. See Theile, *Palitzsch*, p. 150, who refers to 'K. A. Engelhard, Denkwürdigkeiten aus der sächsischen Geschichte Band 1, S. 312'. This presumably is: Engelhardt, *Tägliche Denkwürdigkeiten*, who, in vol. 1 on pp. 312–16, narrates many fictitious details about Palitzsch, nevertheless this particular one appears to be missing.

103 Schöne, 'Die Leipziger Ökonomische Sozietat', p. 78. The membership list for 1770 (*Leipziger Adreß-, Post- und Reise-Kalender* (1770), pp. 69–83) includes a few gardeners and mechanicians, but no other farmer.

104 Sächsisches Staatsarchiv, Staatsfilialarchiv Bautzen, 50146 Gutsherrschaft Jänkendorf, no. 79.

105 SLUB, Mscr.Dresd.App.26,298, http://digital.slub-dresden.de/id495978175 [accessed 29 January 2023]: letter by Friedrich Wilhelm Ferber to Christian Gottfried Schulze, 14 July 1779. It seems likely that Ferber had also commissioned the portrait by Graff, painted in oil, as Theile,

Illustration 8. Engraving of Palitzsch's 'head' by C.G. Schulze in 1782 after a portrait in Ferber's possession by the acclaimed court painter Anton Graff. Photo by Andreas Diesend (*Collection Staatliche Kunstsammlungen Dresden, Kupferstich-Kabinett, A 1995-10480*)

Palitzsch, pp. 163–64 assumed. According to Roland Enke, curator for German Painting at the Gemäldegalerie Alte Meister in Dresden, this portrait, which is now kept at the Mathematisch-Physikalischer Salon in Dresden (inv. B VIII 24 01), might have been finished around 1777/78 since it was exhibited in the spring of 1778 at the annual exhibition of the *Kunstakademie* (art academy) in Dresden (unpublished expertise on the painting by Graff).

time, Schulze was in Paris studying with the renowned Johann Georg Wille (1715–1808). From the beginning, Ferber intended a 'national piece' (*Nationalstück*) and gave detailed instructions regarding the engraving's format and composition.[106] Palitzsch's portrait was to be engraved according to Graff's painting, which, therefore, was sent to Paris. It was accompanied by a drawing by Johann Eleazar Schenau (1737–1806), who knew Wille personally and, since 1776, has been one of the two directors of the Dresden Art Academy. Schenau's drawing added the portrait's framing. It showed agricultural and astronomical attributes such as grain ears, a globe and telescope, and the zodiacal ring. Furthermore, an older engraving by Michael Keyl (1722–1798) was sent, from which Schulze was to take the celestial map displaying Halley's comet. Lastly, Ferber added the inscription, purposefully in Latin, in order to be understood by scholars in Germany as well as in France and perhaps even farther afield:

> Johann Georg Palitzsch, farmer in Prohlis near Dresden, a zealous cultivator of his father's field, excellent astronomer, physicist and botanist, almost a stranger to no science, and all this without a teacher only through his own striving, honest and immaculate, a philosopher in his whole life. Born 11 June 1723, produced at the instigation of his friend F. G. v. F.[107]

Schulze's engraving (Illustration 8), which appeared in 1783, was well-received and Ferber's goal to bring honour 'to our nation' apparently achieved.[108] Ferber's patriotism, as displayed in these short remarks, corresponds with the overall patriotic spirit of the Enlightenment era, however, Palitzsch's merits must have been personally pleasing to him. In 1765, Ferber had authored a treatise that unsparingly diagnosed the devastated state of the Saxon economy and administration after the end of the Seven Years' War and detailed the ways to overcome this situation.[109] This memorandum was

106 SLUB, Mscr.Dresd.App.26,297, http://digital.slub-dresden.de/id495978175 [accessed 29 January 2023]: letter by Friedrich Wilhelm Ferber to Christian Gottfried Schulze, 29 August 1779. Keyl, in 1767, produced his engraving after a visit to Palitzsch: Keller, *Nachrichten*, p. 80.
107 English translation after the German translation by Theile, *Palitzsch*, p. 164. The page with the inscription sent by Ferber does not seem to have survived since it is not contained in the quoted volume of letters. The Latin version on the engraving reads: 'JOANNES GEORGIUS PALIZSCH. Colonus Prohlicii prope Dresdam arvi paterni cultor solertissimus, Astronomus, Physicus, Botanicus egregius; in nullâ fere doctrinâ hospes, ΑΥΤΟΔΙΔΑΚΤΟϹ, probus, candidus, in omni vitâ Philosophus. Natus die XI. Junii, M. DCC. XXIII. FAC. CURAVIT, AMICUS F. G. DE F'.
108 Cf. *Neue Bibliothek der schönen Wissenschaften und der freyen Künste* 29 (1783), pp. 323–24. For Ferber see: SLUB, Mscr.Dresd.App.26,298.
109 *L'esprit et le système du gouvernement de la Saxe depuis la mort du feu roi Auguste III. jusqu'a l'année 1765*. The text is signed 'Dresde le 12 de Juin 1765', but appeared first in print in 1784. The

sent to the courts of Europe. In issuing Palitzsch's image, Ferber conveyed another message: one of success. Thus, Palitzsch's connection to men like Ferber, profoundly influenced by the ideas of the Enlightenment, shaped his image – in life and posthumously – as much as it shaped the overall ideal of the educated commoner. The acclaimed 'peasant', however, also worked on his self-presentation.

The Public Image of Science and the Public Image of the Science Enthusiast

Both Gärtner and Palitzsch made use of opportunities for publicity. Gärtner first appeared in the local papers in September 1757 after he had observed a comet (C/1757 R1). This news was published by the Dresden-based *Historischer Kern Dresdnischer Merkwürdigkeiten*, edited by Johann Christian Crell (1690–1762) and by the far-reaching *Leipziger Zeitungen*, edited by Johann Heinrich Liebers (1700–1764).[110] Regarding this first appearance, we can only speculate how the papers caught wind, however, both editors were certainly well connected.[111] While the *Historischer Kern Dresdnischer Merkwürdigkeiten* generally adopted a short reporting style, the *Leipziger Zeitungen* changed from the first heading 'Astronomical Report That Has Come Our Way' (*Astronomischer Bericht, der an uns gelanget ist*) to referring to letters received from Gärtner, who thus at a certain point in time commenced to inform the editor himself.[112] Palitzsch did likewise, perhaps emulating Gärtner. Palitzsch's reports, for instance, on the occasion of the comet of 1760 (C/1760 A1), were more detailed than Gärtner's as they included general reflections and more specified information on the comet's position.

manuscript, which according to Schlechte, *Die Staatsreform*, p. 61, n. 246 is kept in the *Sächsische Staatsarchiv Dresden*, was sent to the courts in Paris and Vienna. Regarding patriotism and the German Enlightenment, see: Böning, 'Das "Volk"', pp. 63–98; Prignitz, *Vaterlandsliebe*, pp. 7–38.

110 In my essay 'Wissenschaft', I name Gottlieb Schumann as editor of the *Leipziger Zeitungen*. Yet Schumann apparently commenced this work in 1762, hence, the editor in question seems to have been Liebers, although references in the secondary literature are ambiguous. See: Hillert, 'Schumann', pp. 369–77 and Von Witzleben, *Geschichte*, pp. 38, 217. Regarding Crell see: Schöne, *Die Anfänge*, pp. 24–27.

111 Liebers, for instance, had studied in Leipzig. In 1727, he became a member of the Deutsche Gesellschaft. Gottsched was its president ('Senior') from 1727 to 1738. Cf. the entry for Liebers in the 'Bio-bibliographisches Korrespondentenverzeichnis der Bände 1 bis 14' for: Döring et al., *Johann Christoph Gottsched*, vols. 1–14 (Berlin 2007–2020). I would like to thank Betty-Maria Dafis from de Gruyter for sending me a pdf of this register.

112 *Leipziger Zeitungen*, 19 September 1757, p. 596 and 31 July 1758, p. 484.

Palitzsch used Bayer's *Uranometria* and Doppelmayer's *Atlas Coelestis* in order to indicate the comet's latitude and longitude relative to the stars of the constellation in which it was seen.[113] Gärtner, on the other hand, described the course of the comet in rather general terms. He positioned it approximately in relation to the constellations, for instance: 'besides the constellation Orion at the bottom next to the shield on the right hand by the circle of the equator'.[114] Although Gärtner's inaccuracy mattered for astronomers such as Klinkenberg, it did not interest the papers. What mattered for the press was the question of priority.

Although differently oriented, the *Historischer Kern Dresdnischer Merkwürdigkeiten*, the *Dresdner Gelehrte Anzeigen*, and the *Leipziger Zeitungen* all published notices on unusual and peculiar natural phenomena, be it severe frost, thunderstorms, remarkably bright northern lights, or eclipses of the Sun and Moon. These notices were usually included by the editor without naming the observers. Yet, the comet sightings of 1757 and 1758 stood out as Gärtner explicitly claimed the discoveries for himself. In so doing, he set the tone that characterised the future handling of the topic: the 'devotees', as these scientifically interested laypeople called themselves, competed in being the most assiduous. Thus Gärtner, although having been beaten by five days in the discovery of Halley's comet, declared in March 1759 that he had been the first to see it coming out of the Sun, i.e. after the comet had reached its perihelion.[115] Palitzsch, on the other hand, reminded his readers in 1760 that he had been the one to reveal Halley's comet to the learned world.[116] Within this competition, every hour mattered. Thus, in 1760, Gärtner announced in the *Leipziger Zeitungen* that he had discovered a comet on 9 January at 7 o'clock in the evening.[117] Yet Palitzsch declared in the *Dresdner Gelehrte Anzeigen* that he had seen the comet the same day at 6 o'clock in the evening.[118] The *Historischer Kern Dresdnischer Merkwürdigkeiten* then added to its report on Palitzsch's observation that Gärtner had 'also discovered' the beautiful comet star.[119] In February of the same year, Palitzsch only came second, while the very first person to see this comet remained

113 *Dresdner Gelehrte Anzeigen*, January 1760, col. 88–89. Bayer, *Uranometria*. Later editions appeared in 1639 and 1661. Doppelmayer, *Atlas Coelestis*.

114 *Historischer Kern Dresdnischer Merkwürdigkeiten* 2 (1760), p. 5: 'Er stehet neben den Sternbilde Orion unten neben dem Schilde rechter Hand bey dem Aequatorzirkel'.

115 *Leipziger Zeitungen*, 28 March 1759, p. 204.

116 *Dresdner Gelehrte Anzeigen* 6 (1760), col. 87–90 (87).

117 *Leipziger Zeitungen*, 15 January 1760, p. 36.

118 *Dresdner Gelehrte Anzeigen* 6 (1760), col. 87–90.

119 *Historischer Kern Dresdnischer Merkwürdigkeiten* 2 (1760), p. 5.

unnamed.[120] In August 1769, things grew particularly bitter for Gärtner. In number XVI the *Historischer Kern Dresdnischer Merkwürdigkeiten* noted once more that Gärtner, 'known for his skill in astronomy', had observed a comet, but corrected this in its next issue by informing its readers that the comet had been seen three days earlier by Palitzsch, who, moreover, had been allowed to present his observations to the Elector.[121] Gärtner tried to regain ground by reacting to the notification of the comet's renewed visibility after its journey around the Sun in number XX of the *Historischer Kern*. Number XXI brought his claim to have observed the comet a full six days earlier after its return than Palitzsch.[122] With this, Gärtner vanished from the newspapers, perhaps because of his increasingly difficult financial situation resulting from Seven Years' War. Palitzsch, on the other hand, remained in the public eye.

In 1761, Palitzsch had published an article about the transit of Venus and, in 1764, about the eclipse of the Sun.[123] In 1779, he observed another comet and duly informed the public.[124] Yet the papers further cemented Palitzsch's fame in communicating the favours bestowed on him by the court. In 1769, the *Historischer Kern* mentioned a visit by the electoral couple, in July 1779 one by the Elector's mother.[125] Shortly afterwards, in September, the journal devoted a paragraph to some valuable gifts comprising telescopes and editions of Carl von Linné's *Systema natura* and George Louis Leclerc de Buffon's *Histoire naturelle*, presented to Palitzsch by the princes Heinrich of Prussia and Leopold of Brunswick-Wolfenbüttel.[126] The original note lined up with the regular chatter from the Dresden court, printed in the *Historischer Kern*, which informed its attentive readers about the princes' every movement. Yet the information changed its significance when it was copied by other papers, such as the *Leipziger Zeitungen* and the *Reichs-Postreuter*,

120 The comet note, signed 'G. G. H'. [Georg Gottlieb Haubold?] appeared in the *Dresdner Gelehrte Anzeigen* 7 (1760), col. 109–10. The *Gelehrte Anzeigen* did not report on this comet again. Palitzsch seems to have informed the *Historischer Kern Dresdnischer Merkwürdigkeiten* about his observations, which were reported in no. 3 (1760), p. 12 with the indication that the comet had been seen before.

121 *Historischer Kern Dresdnischer Merkwürdigkeiten* 16 (1769), p. 63; 17 (1769), pp. 65–66.

122 *Historischer Kern Dresdnischer Merkwürdigkeiten* 20 (1769), p. 79; 21 (1769), p. 84. Palitzsch is not named in no. 20; that he had sent in his observations becomes clear in no. 21.

123 *Dresdner Gelehrte Anzeigen* 33 (1761). col. 525–28; 18 (1764), col. 169–74; *Historischer Kern Dresdnischer Merkwürdigkeiten* 7 (1764), p. 26.

124 *Dresdner Gelehrte Anzeigen* 5 (1779), col. 63–64; *Historischer Kern Dresdnischer Merkwürdigkeiten* 2 (1779), p. 8.

125 *Historischer Kern Dresdnischer Merkwürdigkeiten* 15 (1769), p. 59; 13 (1779), pp. 50–51.

126 *Historischer Kern Dresdnischer Merkwürdigkeiten* 14 (1779), p. 55.

which was based in Altona near Hamburg.[127] In addition to the fact that the prince's generosity and Palitzsch's worthiness reached a far wider audience, the announcement acquired a particular importance when placed alongside international news instead of local trivia. In this way, Palitzsch proved to be considerably more successful in the long run than Gärtner. Nonetheless, both Gärtner and Palitzsch employed the same strategy: they offered their astronomical observations as a commodity to the papers and gained a certain public standing in return. This did not, however, remain unchallenged.

Christian Gotthold Hoffmann, already mentioned on several occasions, had issues with Gärtner's reputation. On the occasion of Palitzsch's sighting of Halley's comet, he declared in the *Dresdner Gelehrte Anzeigen*:

> The yarn merchant, Christian Gärtner, at Dolckewitz, who has an indescribable delight in astronomy, but no thorough knowledge of it, but only knows to talk about it in an empirical and obscure manner and only to the uninformed, had had the chance to discover the first two [comets, the one of 1757 and the one of summer 1758].[128]

Palitzsch, in contrast, stood far above Gärtner's ignorant dilettantism 'as he had not only read the relevant books with diligence, but had also learned to thoroughly understand and use them'. Furthermore, Palitzsch had mastered planar and spherical trigonometry and, therefore, possessed theoretical knowledge. Furthermore, he had studied the philosophy of Christian Wolff and, last but not least, enthused over physics and botany.[129] With his characterisation, Hoffmann established a sharp divide between the mere 'stargazer' (*Sterngucker*) Gärtner and the serious amateur Palitzsch,

127 *Leipziger Zeitungen*, 17 July 1779, p. 623; *Reichs-Postreuter* 57 (1779), p. 1 (26 July 1779).

128 *Dresdner Gelehrte Anzeigen* 2 (1759), col. 17: 'Der Zwirnhändler, Christian Gärtner, zu Dolckewiz, welcher zwar eine unbeschreibliche Lust an der Astronomie, aber gar keine gründliche Känntniß in selbiger hat, sondern lediglich nur empyrisch und nur gegen Unkundige etwas dunkles davon zu reden weiß, hat die zwey ersten [comets of 1757 and summer 1758] zu entdecken den Zufall gehabt'.

129 *Dresdner Gelehrte Anzeigen* 2 (1759), col. 17: 'alleine er [Palitzsch] hat sich in der Stern-Wissenschafft noch weit über bemeldeten Sterngucker [Gärtner], hinausgesetzet, indem er nicht nur dergleichen Bücher mit Fleiß gelesen, sondern auch gründlich zu verstehen, und deutlich anzuwenden gelernet, auch die plane und sphaerische Trigonometrie fertig, und in allen Fällen zu gebrauchen weiß, folglich eine theoretische Känntniß besitzet, welche er mit seinen fleißigen Beobachtungen über seinen Stand und die Vermuthung von ihm verbessert, hiernächst aber auch in der Wolffischen Philosophie sich sehr bekannt gemacht, besonders aber die Physic und Botanic zu seinen favorit Wissenschafften erlesen'.

who perfectly complied with the emerging ideal of the educated 'peasant' or countryman and presented himself accordingly.[130]

In the course of the eighteenth century, the traditionally negative image of the peasantry as coarse and ignorant slowly changed into one that ascribed to it an eminent role for the general prosperity and development of society. Representatives of the German Enlightenment began to take an interest in the third estate and the steadfast Wolffian Hoffmann actually presented for the first time a living member of the peasantry in print in his treatise *The Erudite Peasant* (*Der gelehrte Bauer*) published in Dresden in 1756.[131] In this treatise, Hoffmann edited and prefaced three essays by Johann Ludewig (1715–1760), a peasant and tax collector in Cossebaude village near Dresden, whom he had encountered in his position as excise taxes councillor. Ludewig had taught himself mathematics and philosophy to such a degree that he brilliantly passed an exam by Johann Heinrich Winckler, acclaimed professor of physics at Leipzig University and fellow of the Royal Society.[132] Hoffmann's edition of Ludewig's writings is, in fact, the result of another examination set for Ludewig by Hoffmann himself. While the latter is full of praise for Ludewig's intellectual accomplishments, he simultaneously reveals the low regard usually afforded by the educated upper classes to the peasantry. Hoffmann is taken aback when looking at Ludewig's 'stern, simple-minded, in a word, his dry peasant face, with shaggy hair', while hearing him talk wisely. Similarly, he cannot believe that Ludewig really is a peasant when seeing him in the usual habit of a scholar: the nightgown.[133] To be sure, Hoffmann's preface is witty and full of sympathy for 'his good peasant' Ludewig. At the same time, however, it sets the bar very high, for Hoffmann is not interested in the peasantry as such, but searches for scholarship in the peasantry – scholarship that complies with academic standards.

Perhaps unsurprisingly, apart from Ludewig, no peasant had yet been able to fulfil Hoffmann's ideal. The others he had met thus far and who had ventured into scholarship (one wonders how many) all lacked something important, be it in precepts, general understanding, or clarity of notions.[134]

130 In view of Hoffmann's wording, I cannot support the idealistic interpretation that Gärtner and Palitzsch were 'ideal-typical recipients of a two-stage educational program'. Cf. Mahlmann-Bauer, 'Mylius', p. 266.

131 Hoffmann's publication has been repeatedly discussed by Holger Böning, in particular, in the latter's edition of *Der gelehrte Bauer*.

132 Hoffmann, *Der gelehrte Bauer*, pp. 172–76.

133 Hoffmann, *Der gelehrte Bauer*, 'Vorbericht', pp. 3v–4r (page numbering stops here).

134 Hoffmann, *Der gelehrte Bauer*, p. 7v.

One such 'peasant' who displayed a serious dearth of proper comprehension was Gärtner. His lack of theoretical knowledge must have discredited him in Hoffmann's eyes, who held 'that to become wise these days, depends only on the will'.[135] Hoffmann's disparagement of Gärtner and praise of Palitzsch had consequences. Gottsched's journal *Das Neueste aus der anmuthigen Gelehrsamkeit* published Hoffmann's text as it had appeared in the *Dresdner Gelehrte Anzeigen*. Hoffmann had probably sent it simultaneously to both papers. The editor of *Das Neueste aus der anmuthigen Gelehrsamkeit*, perhaps Gottsched himself, who only one year before had spoken highly of Gärtner's jovilabium, added a postscript announcing a forthcoming publication on the comet, but did not comment on Hoffmann's characteristics of Gärtner, which thus circulated unquestioned in Dresden, Leipzig, and beyond.[136] Furthermore, Gärtner's lack of theoretical understanding was communicated internationally due to Heinsius. The latter had been approached by Christian Mayer on behalf of Joseph-Nicolas Deslisle, who sought to obtain more detailed information about the Saxon comet observations, but was to learn that Gärtner was just a peasant without 'any theory'.[137] Heinsius's words made their rounds in Paris and were finally published in the *Histoire de l'Académie Royale des Sciences* for the year 1759, printed in 1765.[138] At this point, however, Heinsius was not concerned with Enlightenment ideas about an educated peasantry. Feeling that his reputation as a serious astronomer was at stake, he aimed at downplaying the impact of amateurs. In line with his deprecating comments on Gärtner, Heinsius presented Palitzsch's discovery of Halley's comet as due to serendipity and imputed that Palitzsch had not recognised the importance of his find.[139]

135 Hoffmann, *Der gelehrte Bauer*, p. 31r: 'daß es heut zu Tage klug zu werden, nur auf das Wollen ankomme'.

136 *Das Neueste aus der anmuthigen Gelehrsamkeit* 1 (1759), p. 72. I could not find information on the circulation of this particular journal. However, Gottsched had numerous and far-reaching contacts. Another one of his journals, the *Neuer Büchersaal der schönen Wissenschaften und freyen Künste* reached all parts of Germany and thus had a broad geographical impact. Both journals were aimed at a general, educated public. Cf. Ball, *Moralische Küsse*, p. 323.

137 See note 92.

138 *Histoire de l'Académie Royale des Sciences. Année MDCCLIX* (1765), p. 142: 'M. Guertner [sic] [...] n'étoit point un Astronome consommé, mais un habitant de la campagne que s'est applicqué à faire quelques observations, sans avoir eu aucune théorie, & qu'il ne povoit donner de position exacte de cette Comète'.

139 Anon. [Heinsius, Gottfried], *Anzeige*. Heinsius admitted his authorship in a letter to Leonhard Euler: Archive of the Russian Academy of Sciences (RAS), Saint Petersburg branch, Ф. 136. Оп. 2. Д. 4. Л. 96, 96об., 97, 97об: Letter from Heinsius to Euler, 22 April 1759. I sincerely thank Martin Mattmüller from the Bernoulli-Euler-Zentrum of the University of Basel for providing me with a copy of this letter. In addition to downplaying Palitzsch's search for the comet, Heinsius did

Commoners and Science in Public: A Mixed Blessing

Gärtner and Palitzsch belonged to the possibly not so small group of common people who did not fulfil the clichés of their social rank. They not only distinguished themselves through the knowledge they acquired but, most of all, through establishing useful networks of contacts. Both were immersed in various local circles, which entailed contacts within a broad social spectrum and could support them in various ways in order to fulfil their scientific ambitions. Regarding astronomy, especially the strong contemporary interest in comets offered humble 'stargazers' such as Gärtner the opportunity to be noticed far beyond the confines of their original station in life. Since the astronomical community strove to observe each appearing comet as long as possible in order to gather sufficient data to determine its path, each sighting became important. The attachment of names added further relevance and reliability. Thus Heinsius, although not caring much for Gärtner's astronomical abilities, dutifully reported to Euler that Gärtner had been the first to see Halley's comet again after its perihelion.[140] Perhaps most striking is the sense for self-publicity that Gärtner and Palitzsch displayed in the use of the papers.

However, the ensuing popularity proved a double-edged sword, as particularly Gärtner was to discover: although he was very proud to have been approached by several academics, he could hardly comply with their requests. Despite isolated interventions such as Kästner's, Hoffmann's influential opinion that Gärtner lacked knowledge and understanding became widely accepted.[141] It was repeated in the *Wittenbergsches Wochenblatt*, edited by the professor of physics, Johann Daniel Titius (1729–1796). Yet the anonymous author, perhaps Titius himself, added: 'that he [Gärtner] was not ignorant in various physical as well as optical and mechanical things, and had himself brought about different celestial notions in machines'.[142] Since Titius and his *Wittenbergsches Wochenblatt* advocated the practical effectiveness of enlightened endeavours,[143] Gärtner's mechanical skills were

not mention Palitzsch's use of an 8-foot telescope. Because of this omission, Delisle came to think that Palitzsch had discovered the comet with the naked eye. Craig B. Waff also inferred that Heinsius had authored the text ('The First International Halley Watch', p. 401).

140 RAS, letter from Heinsius to Euler, 22 April 1759.

141 *Allgemeine deutsche Bibliothek* 30, no. 1 (1777), p. 264. I would like to thank Friedhelm Schwemin for bringing Kästner's authorship to my attention.

142 Anon. [Titius?], 'Nachricht', 51. St., 1782, p. 402, 'er in unterschiedlichen physischen auch optischen und mechanischen Dingen nicht unkundig war, und selbst unterschiedliche himmlische Vorstellungen in Maschinen zu Stande gebracht hatte'.

143 For Titius and the idea of practical enlightenment, see: Böning, 'Das Intelligenzblatt', pp. 118–29.

singled out and resulted in some recognition. But they were of no influence on the general ideal of the educated peasant.

Palitzsch, on the other hand, seemed the living embodiment of this ideal. His finances, collections, and leaning smoothed the rusticity of the peasant and made him fit perfectly in the circle of erudite gentlemen, within which he moved. According to his friend Karl Wilhelm Dassdorf, one of Palitzsch's greatest merits besides all his accomplishments lay in the fact that he never forgot 'the duties of this sphere of activity in which Providence had placed him'.[144] His example thus served to counteract the prominent contemporary fear, voiced on many occasions in the *Schattenrisse*, that uncommon, unbefitting interests would lead to a neglect of breadwinning employment.[145] Palitzsch himself shared this view. In a letter to the Elector, written in 1776, asking to be relieved of work obligations in one of the Elector's agricultural estates (*Vorwerk*), Palitzsch declared:

> From the first years of my youth, driven by a natural inclination to investigate and to acquire all kinds of useful knowledge, I believed to find the promotion of true happiness in the connection of the unceasing continuation of rural economy with occupations, to which my status and my way of life did not lead me just as directly; at the same time, in the continuing endeavour to be able to judge correctly of my condition and conduct, I have daily convinced myself more and more of the truth of the principle that every man is especially bound, in his occupations and actions, to exert himself to the utmost in the fulfilment of those duties which his state and profession, assigned to him by Providence, demand of him. Although my extraordinary love for the sciences, in particular astronomy, natural history and the contemplation of God's works in general, often made me tempted to occupy myself less than others of my estate with agriculture, and to avoid the hardships of my way of life to a certain extent, this notion has triumphed, as I now praise God daily.[146]

144 Dassdorf, *Beschreibung*, p. 780.

145 [Lose], *Schattenrisse*, vol. 3, pp. 134, 136, 226; Böning, 'Die Entdeckung des "Volkes"', pp. 276–78.

146 SachStA-D, 10047 Amt Dresden, Nr. 915, fol. 3r–v: 'Von erstern Jahren meiner Jugend, von natürlicher Neigung, zum Nachforschen und zur Erwerbung aller Gattung nützlicher Kenntniße, angetrieben, in der Verbindung der unabläßigen Fortsetzung der LandWirthschaft mit Beschäftigungen, zu welchen mein Stand und meine LebensArt mich nicht eben so unmittelbar geführet, der Beförderung einer wahren Glückseeligkeit zu finden geglaubet; so habe ich zu gleich, beÿ dem anhaltenden Bestreben, über meinen Zustand und Verhaltens richtig urtheilen zu können, mich täglich mehr von der Wahrheit des Grundsatzes überzeuget, daß ein jeder Mensch, besonders verbunden seÿ, beÿ seinen Beschäftigungen und Handlungen, auf die Erfüllung derjenigen Pflichten, welche ihm der von der Vorsehung angewiesene Stand und Beruff,

In this, almost programmatic, manner Palitzsch lived the ideals of the Enlightenment: his exemplary conduct brought him prosperity – and simultaneously banished fears of social unrest.[147] Gärtner's unhappy history, by contrast, played into the hands of the critics of popular Enlightenment. His miserable end in poverty, illness, and isolation, described by the *Schattenrisse*, could be interpreted as a natural consequence of his neglect of his original profession, while his unsolicited ambitions and efforts gained him nothing 'but the shining reputation of a learned peasant, which eventually made him a beggar'.[148] Palitzsch, in contrast, perfectly fulfilled the image of the educated commoner as defined by the enlightened bourgeoisie: knowledgeable, particularly in the theoretical side of the sciences, versed in many ways, but still humble, dutiful, and content in his station.

The contemporary papers did their part to disseminate this image. After his death in 1788, a remarkable number of announcements and obituaries appeared, in local, national, and international newspapers. Most of them praised Palitzsch as a man 'in the old Roman spirit', who worked his land with his own hands, lived as a philosopher, maintained contacts to heroes and princes, was a permanent correspondent of the London Royal Society, and had discovered a comet in 1738![149] As in the case of Gärtner's comet news, the papers followed the contemporary habit of copying and extracting from one another. Thus, the *Deutsche Zeitung*, the *Der Bote aus Thüringen*, the *Magazin für das Neueste aus der Physik und Naturgeschichte*, the *Ommelander Courant*, *Stockholms Post-Tidningar*, the *Mercure de France*, and *The Gentleman's Magazine* and probably others all repeated the mistake originating in the *Leipziger Zeitungen*, which had erroneously printed the

auflegen, vorzüglichen Eifer zu wenden. Ohngeachtet als die außerordentliche Liebe zu denen Wißenschaften, insonderheit der Astronomie, der Naturkunde und der sorgfältigen Betrachtung der Werke Gottes überhaupt, mich oft der Versuchung ausgesezet, weniger, als andere meines Standes, mit der Landwirthschaft mich zu beschäftigen, und den Beschwerlichkeiten meiner Lebensart mich einigermaßen zu entziehen; so hat jedoch wie ich Gott nunmehro täglich preise, jene Betrachtung gesieget'.

147 Cf., for instance, Schaffer, 'Natural Philosophy'.

148 [Lose], *Schattenrisse*, vol. 3, p. 134: 'Dafür erhielt der unermüdete gefällige Mann [Gärtner] denn weiter nichts, als die glänzende Reputation eines gelehrten Bauers, die ihn endlich zum Bettler machte'.

149 *Leipziger Zeitungen*, 3 March 1788, p. 259: 'Er [Palitzsch] war ein Mann im alten Römischen Geiste, der mit eignen Händen den väterlichen Acker bauete, als ein Weiser lebte, und mit vielen Talenten und Kenntnissen auch Menschenfreundlichkeit und edle Sitten verband. Er war einer der fleißigsten astronomischen Beobachter. Im Jahr 1738 entdeckte er zuerst einen Kometen, und wurde deswegen von der Societät der Wissenschaften in London mit vielem Lobe beehrt; so wie er denn auch ihr beständiger Correspondent war'. In the *Dresdner Gelehrte Anzeigen* an obituary appeared in number 11 (1788) with a different text.

year 1738 instead of 1758.[150] Furthermore, Palitzsch had never been a corresponding member of the Royal Society, which simply did not have this membership category. While the sheer number of papers and journals that found Palitzsch's death noteworthy remains a remarkable fact in itself, it is evident that the media as much as the *Schattenrisse* lauded an ideal rather than an individual.

The cases of Gärtner and Palitzsch thus illuminate the interplay of different factors such as curiosity, resourcefulness, and networks of contacts that, beside the point of actual knowledge, contributed to the public visibility of these protagonists, affected their reputation, and determined the successfulness of their efforts. The available sources hardly make it possible to distinguish between the level of knowledge Gärtner and Palitzsch had each achieved in the manner of their eighteenth-century contemporaries. In relation to the professional astronomy of their time, whose main concern was to sample proof for the Newtonian world system and thus needed exact observational data, both were of negligible consequence. Yet their stories show the interest and inquisitiveness in the natural sciences that could be found in the so-called lower classes, as much as the impact and connotations such knowledge held in an increasingly differentiated social environment.

Bibliography

Achermann, E., ed., *Johann Christoph Gottsched (1700–1766): Philosophie, Poetik und Wissenschaft* (Berlin 2014).

Anon. [Heinsius, Gottfried], *Anzeige daß der im Jahre 1682 erschienene und von Halley nach der Newtonianischen Theorie auf gegenwärtige Zeit Vorherverkündigte Comet wirklich sichtbar sey; und was derselbe in der Folge der Zeit für Erscheinungen haben werde* (Leipzig 1759).

Anon. [Titius?], 'Nachricht von einem gelehrten Landmann im Vogtländischen', in: *Wittenbergsches Wochenblatt zum Aufnehmen der Naturkunde und des ökonomischen Gewerbes*, 51. St., 1782, pp. 401–5.

'Anzeige von der neu errichteten Leipziger ökonomischen Societät', *Leipziger Intelligenz Blatt* 24 (1764): *Beylage zum 24sten Stück*, pp. 217–20.

150 *Deutsche Zeitung* 12 (1788), p. 94; *Der Bote aus Thüringen* 13 (1788), p. 208; *Magazin für das Neueste aus der Physik und Naturgeschichte* 5, no. 3 (1788), pp. 178–79; *Ommelander Courant* 26 (1788), p. 2; *Stockholms Post-Tidningar* 27 (1788), p. 2; *Mercure de France*, 29 March 1788, p. 205. *The Gentleman's Magazine* (April 1788, p. 366) further distorted the incorrect year into 1783. I wish to express my sincerest gratitude to Robert Heinz from the Palitzsch-Museum in Dresden, who provided me with various obituaries.

Baasner, R., *Das Lob der Sternkunst: Astronomie in der deutschen Aufklärung* (Göttingen 1987).
Baasner, R., *Abraham Gotthelf Kästner, Aufklärer: (1719–1800)* (Berlin/Boston 1991).
Ball, G., *Moralische Küsse: Gottsched als Zeitschriftenherausgeber und literarischer Vermittler* (Göttingen 2000).
Bayer, J., *Uranometria* (Augsburg 1603).
Bensaude-Vincent, B., and C. Blondel, ed., *Science and Spectacle in the European Enlightenment* (Aldershot 2008).
Böning, H., 'Das Intelligenzblatt als Medium praktischer Aufklärung: Ein Beitrag zur Geschichte der gemeinnützig-ökonomischen Presse in Deutschland von 1768 bis 1780', *Internationales Archiv für Sozialgeschichte der deutschen Literatur* 12 (1987), pp. 107–33.
Böning, H., 'Das "Volk" im Patriotismus der deutschen Aufklärung', in: O. Dann et al., ed., *Patriotismus und Nationsbildung am Ende des Heiligen Römischen Reiches* (Cologne 2003), pp. 63-98
Böning, H., 'Die Entdeckung des "Volkes" in der deutschen Aufklärung', in: J. Ludewig, *Der gelehrte Bauer: Mit D. Christian Gotthold Hoffmanns Vorbericht nebst Kupffern, Neudruck der ersten Auflage Dresden 1756, mit einem Nachwort von Holger Böning* (Stuttgart 1992), pp. 253–82.
Böning, H., 'Popularaufklärung – Volksaufklärung', in: R. van Dülmen and S. Rauschenbach, ed., *Macht des Wissens: Die Entstehung der modernen Wissensgesellschaft* (Cologne 2004), pp. 563–81.
Böning, H., 'Die Entdeckung des gemeinen Mannes in der Aufklärung: Philosophische, gelehrte und literarische Bauern in der deutschen Öffentlichkeit', in: H. Böning et al., ed., *Selbstlesen – Selbstdenken – Selbstschreiben: Prozesse der Selbstbildung von 'Autodidakten' unter dem Einfluss von Aufklärung und Volksaufklärung vom 17. bis zum 19. Jahrhundert* (Bremen 2015), pp. 117–64.
Bruhns, C., 'Heinsius, Gottfried', in: *Allgemeine Deutsche Biographie* 11 (1880), p. 656. www.deutsche-biographie.de/sfz29360.html [accessed 29 April 2024].
Dassdorf, K.W., ed., *Winckelmanns Briefe an seine Freunde*, vols. 1–2 (Dresden 1777, 1780).
Dassdorf, K.W., *Beschreibung der vorzüglichsten Merkwürdigkeiten der Churfürstlichen Residenzstadt Dresden und einiger seiner umliegenden Gegenden* (Dresden 1782).
Daum, A.W., 'Varieties of Popular Science and the Transformations of Public Knowledge: Some Historical Reflections', *Isis* 100, no. 2 (2009), pp. 319–32.
De Fontenelle, B., *Entretiens* (Paris 1686) [transl. Gottsched, J.C., *Gespräche von mehr als einer Welt* (Leipzig 1726)].
Doppelmayer, J.G., *Atlas Coelestis* (Nuremberg 1742).
Döring, D., *Johann Christoph Gottsched in Leipzig* (Leipzig 2000).

Döring, D. et al., ed., *Johann Christoph Gottsched: Briefwechsel unter Einschluss des Briefwechsels von Luise Adelgunde Victorie Gottsched*, vols. 1–14 (Berlin 2007–2020).

Duhr, B., *Geschichte der Jesuiten in den Ländern deutscher Zunge*, vols. 1–4 (Munich 1907–1928).

Ebert, F.A., *Briefwechsel Christian Fürchtegott Gellert's mit Demoiselle Lucius* (Leipzig 1823).

Engelhardt, K.A., *Tägliche Denkwürdigkeiten aus der sächsischen Geschichte*, vols. 1–3 (Dresden 1809–1812).

Ferber, F.W., *L'esprit et le système du gouvernement de la Saxe depuis la mort du feu roi Auguste III. jusqu'a l'année 1765* (s.l. 1784).

Fischer, W., *Mineralogie in Sachsen von Agricola bis Werner: Die ältere Geschichte des Staatlichen Museums für Mineralogie und Geologie zu Dresden (1560–1820)* (Dresden 1939).

Gluch, S., 'Wissenschaft als Mittel sozialer Distinktion: Christian Gärtner und seine Kontakte zur gelehrten Astronomie des 18. Jahrhunderts', in: W.R. Dick and J. Hamel, ed., *Beiträge zur Astronomiegeschichte*, vol. 14 (Leipzig 2019), pp. 37–65.

Golinski, J., *British Weather and the Climate of Enlightenment* (Chicago 2007).

Gottsched, J.C., *Historische Lobschrift des weiland hoch- und wohlgebohrnen Herrn Herrn Christians, des H. R. R. Freyherrn von Wolf* (Halle 1755).

Gottsched, J.C., 'Nachricht und Beschreibung einer neuerfundenen Maschine, den Lauf der Trabanten Jupiters vorzustellen', *Das Neueste aus der anmuthigen Gelehrsamkeit* 3 (1758), pp. 229–32.

Gruber, D., 'Die Kometenerscheinung von 1743/44 in Gelehrten Journalen:. Eine Annäherung', in: K. Löffler, ed., *Wissen in Bewegung: Gelehrte Journale, Debatten und der Buchhandel der Aufklärung* (Stuttgart 2020), pp. 157–69.

Gruber, D., *Frühneuzeitlicher Wissenswandel: Kometenerscheinungen in der Druckpublizistik des Heiligen Römischen Reiches* (Bremen 2020).

Günther, H.P., *Die Casanova Tour: Mr Nugent's Wechselkurse*. www.giacomo-casanova.de/geld.htm [accessed 29 April 2024].

Hell, M., *Observatio transitus Veneris ante discum Solis die 5ta Junii 1761* (Vienna [1761]).

Hell, M., *Ephemerides astronomicae anni bisexti 1764 ad meridianum vindobonensem* (Vienna 1763).

Hell, M., *Ephemerides astronomicae anni 1765 ad meridianum vindobonensem* (Vienna 1764).

Hell, M., *Ephemerides astronomicae anni bisexti 1768 ad meridianum vindobonensem* (Vienna 1767).

Herbst, K.-D., ed., *Astronomie, Literatur, Volksaufklärung: Der Schreibkalender der Frühen Neuzeit mit seinen Text- und Bildbeigaben* (Bremen 2012).

Herbst, K.-D., 'Bäuerliche Autodidakten als Astronomen und Kalendermacher', in: H. Böning et al., ed., *Selbstlesen – Selbstdenken – Selbstschreiben: Prozesse der Selbstbildung von 'Autodidakten' unter dem Einfluss von Aufklärung und Volksaufklärung vom 17. bis zum 19. Jahrhundert* (Bremen 2015), pp. 175–90.

Hillert, S., 'Gottlieb Schumann – Redakteur der "Leipziger Zeitungen" während der Studienjahre Radiščevs in Leipzig', *Wissenschaftliche Zeitschrift der Karl-Marx-Universität Leipzig, Gesellschafts- und Sprachwissenschaftliche Reihe* 26 (1977), pp. 369–77.

Histoire de l'Académie Royale des Sciences. Année MDCCLIX (1765).

Hoffmann, C.G., ed., *Der gelehrte Bauer* (Dresden 1756).

Keller, H., *Nachrichten von allen in Dresden lebenden Künstlern* (Leipzig 1788).

Keller, K., 'Saxony: *Rétablissement* and Enlightened Absolutism', *German History* 20, no. 3 (2002), pp. 309–31.

Kindermann, E.C., *Vollständige Astronomie, Oder: Sonderbare Betrachtungen derer vornehmsten an dem Firmament befindlichen Planeten und Sternen* [...] (Rudolstadt 1744).

Kindermann, E.C., *Collegium astronomicum, als der andere Theil oder Erklärung seiner teutschen Astronomie* (Dresden 1747).

[Lose], *Schattenrisse edler Teutschen: Aus dem Tagebuche eines physiognomischen Reisenden*, vols. 1–3 (Halle 1783–1784).

Ludewig, J., 'Kurtze Berechnung und Abriß der am 26. Oct. 1753. sich ereignenden Sonnen-Finsterniß', in: C.G. Hoffmann, ed., *Der gelehrte Bauer* (Dresden 1756), pp. 126–45.

Mahlmann-Bauer, B., 'Christlob Mylius, der Wissenschaftsjournalist', in: M. Mulsow et al., ed., *Aufklärung: Interdisziplinäres Jahrbuch zur Erforschung des 18. Jahrhunderts und seiner Wirkungsgeschichte* 31 (2019), pp. 259–96.

Matzerath, J., '"Pflicht ohne Eigennutz" Das kursächsische *Rétablissement*: Restauration einer Ständegesellschaft', *Neues Archiv für sächsische Geschichte* 66 (1995), pp. 157–82.

McConnell, A., *Jesse Ramsden (1735–1800): London's Leading Scientific Instrument Maker* (Aldershot 2007).

Mühlpfordt, G., 'Zwischen Tradition und Innovation: Rektoren der Universität Leipzig im Zeitalter der Aufklärung', in: H. Marti and D. Döring, ed., *Die Universität Leipzig und ihr gelehrtes Umfeld 1680–1780* (Basel 2004), pp. 111–94.

Mulsow, M., *Freigeister im Gottsched-Kreis: Wolffianismus, studentische Aktivitäten und Religionskritik in Leipzig 1740–1745* (Göttingen 2007).

Nitzschke, K., 'Canzler, Karl Christian', in: *Sächsische Biographie*. https://saebi.isgv.de/biografie/Karl_Christian_Kanzler_(1733–1786) [accessed 29 April 2024].

Nitzschke, K. 'Dassdorf, Karl, Wilhelm', in: *Sächsische Biografie*. https://saebi.isgv.de/biografie/Karl_Wilhelm_Dassdorf_(1750–1812) [accessed 29 April 2024].

Palitzsch, J.G., 'Von der Venus ihrer Erscheinung in der Sonnen-Scheibe, am 6. Junii 1761', *Dresdner Gelehrte Anzeigen* (1761), col. 525–28.

Pescheck, C., *Vorhof der Stern-Wissenschafft Oder Astronomiae, Darinnen Alles dasjenige, was von dieser nützlichen Scientz, einem politen Menschen und Anfänger der Astronomie, zu wissen nöthig, deutlich erkläret und beschrieben wird* [...] (Bautzen 1729).

Pfarrerbuch Sachsen. https://pfarrerbuch.de/sachsen/pfarrer [accessed 29 April 2024].

Prignitz, C., *Vaterlandsliebe und Freiheit: Deutscher Patriotismus von 1750 bis 1850* (Wiesbaden 1981).

Schaffer, S., 'Natural Philosophy and Public Spectacle in the Eighteenth Century', *History of Science* 21 (1983), pp. 1–43.

Schaffer, S., 'Halley, Delisle, and the Making of the Comet', in: N.J.W. Thrower, ed., *Standing on the Shoulders of Giants: A Longer View of Newton and Halley* (Berkeley 1990), pp. 254–98.

Schatzberg, W., 'Gottsched as a Popularizer of Science', *MLN* (*German Issue*) 83, no. 5 (1968), pp. 752–70.

Schillinger, K. 'Johann Gottlob Rudolph (1721–1776) – Lebensweg eines bedeutenden sächsischen Mechanikus', *Feingerätetechnik* 7 (1986), pp. 319–21.

Schillinger, K., 'Joachim Friedrich Meyen (1704–1772) – Instrumentenmacher und – verkäufer sowie Inspektor des Mathematisch-Physikalischen Salons', in: J. Hamel and M. Korey, ed., *Weiter sehen: Beiträge zur Frühgeschichte des Fernrohrs und zur Wissenschaftsgeschichte Augsburgs in memoriam Inge Keil* (Frankfurt a.M. 2012), pp. 203–28.

Schlechte, H., *Die Staatsreform in Kursachsen 1762–1763: Quellen zum kursächsischen Rétablissement nach dem siebenjährigen Kriege* (Berlin 1958).

Schöne, A., 'Die Leipziger Ökonomische Sozietät von 1764 bis 1825', *Neues Archiv für sächsische Geschichte* 70 (1999 [2000]), pp. 53–78.

Schöne, W., *Die Anfänge des Dresdner Zeitungswesens im 18. Jahrhundert* (Dresden 1912).

Schröter, J.S., *Journal für die Liebhaber des Steinreichs und der Konchyliologie* 5 (Weimar 1779), p. 170.

Siegert, R., 'Der "gemeine Mann" und die Welt der Bücher um 1800', *Jahrbuch für Kommunikationsgeschichte* 4 (2002), pp. 32–51.

Sorel, P., 'Die Verbreitung und Popularisierung naturwissenschaftlichen und technologischen Wissens durch die Messagers boiteux in der Zeit von 1789 bis 1848', in: S. Greilich and Y.-G. Mix, ed., *Populäre Kalender im vorindustriellen Europa: Der 'Hinkende Bote' / 'Messager boiteux'. Kulturwissenschaftliche Analysen und bibliographisches Repertorium. Ein Handbuch* (Berlin 2006), pp. 202–27.

Theile, F., *Johann Georg Palitzsch. Ein Lebensbild von Dr. Friedrich Theile* (Leipzig 1878).

Van Musschenbroek, P., *Elementa Physica* (Leiden 1734) [transl. J.C. Gottsched, *Hrn. Peters von Muschenbroek M. D. der Weltw. und Mathem. ordentlichen Lehrers zu Leyden, Grundlehren der Naturwissenschaft* (Leipzig 1747)].

Von Witzleben, C.D., *Geschichte der Leipziger Zeitung: Zur Erinnerung an das zweihundertjährige Bestehen der Zeitung* (Leipzig 1860).

Waff, C.B., 'The First International Halley Watch: Guiding the Worldwide Search for Comet Halley, 1755–1759', in: N.J.W. Thrower, ed., *Standing on the Shoulders of Giants: A Longer View of Newton and Halley* (Berkeley 1990), pp. 373–411.

Walters, A.N. 'Conversation Pieces: Science and Politeness in Eighteenth-Century England', *History of Science* 35 (1997), pp. 121–54.

Walters, A.N., 'Ephemeral Events: English Broadsides of Early Eighteenth-Century Solar Eclipses', *History of Science* 37 (1999), pp. 1–43.

Wölfel, K., 'Gellert, Christian Fürchtegott', in: *Allgemeine Deutsche Biographie* 6 (1964), pp. 174–75. www.deutsche-biographie.de/sfz20280.html [accessed 29 April 2024].

Zuidervaart, H.J., *Van 'konstgenoten' en hemelse fenomenen: Nederlandse sterrenkunde in de achttiende eeuw* (Rotterdam 1999).

Journals and Periodicals

Allgemeine deutsche Bibliothek

Das Neueste aus der anmuthigen Gelehrsamkeit

Der Bote aus Thüringen

Deutsche Zeitung

Dresdner Gelehrte Anzeigen

Dresdnisches Magazin, oder Ausarbeitungen und Nachrichten, zum Behuf der Naturlehre, der Arzneykunst, der Sitten und der schönen Wissenschaften

Für Aeltere Litteratur und Neuere Lectüre

Gazette de Bruxelles

Göttingische Zeitung von gelehrten Sachen

Hamburgisches Magazin, oder gesammlete Schriften, aus der Naturforschung und den angenehmen Wissenschaften überhaupt. Continued as: *Neues Hamburgisches Magazin oder Fortsetzung gesammleter Schriften, aus der Naturforschung, der allgemeinen Stadt- und Land-Oekonomie, und den angenehmen Wissenschaften überhaupt*

Historischer Kern Dresdnischer Merkwürdigkeiten

Leipziger Adreß-, Post- und Reise-Kalender

Leipziger Intelligenz-Blatt

Leipziger Zeitungen

Magazin für das Neueste aus der Physik und Naturgeschichte

Mercure de France

Neue Bibliothek der schönen Wissenschaften und der freyen Künste
Neue Versuche nützlicher Sammlungen, zu der Natur- und Kunst-Geschichte, sonderlich von Ober-Sachsen
Neuer Büchersaal der schönen Wissenschaften und freyen Künste
Ommelander Courant
Philosophical Transactions
Reichs-Postreuter
Stockholms Post-Tidningar
The Gentleman's Magazine
Wittenbergsches Wochenblatt

Consulted Archives and Libraries
Archive of the Russian Academy of Sciences (RAS)
Archives nationales, Paris
Bibliothèque de l'observatoire de Paris (BOP)
Niedersächsisches Landesarchiv Wolfenbüttel (NLA WO)
Noord-Hollands Archief, Haarlem, Koninklijk Instituut der Wetenschappen
Sächsische Landes- und Universitätsbibliothek (SLUB)
Sächsisches Staatsarchiv, Hauptstaatsarchiv Dresden (SächsStA-D)
Sächsisches Staatsarchiv, Staatsfilialarchiv Bautzen
University Archive Vienna

About the Author

Sibylle Gluch is director of the Erzgebirge Toy Museum and Open-Air Museum Kurort Seiffen.

6. In Service to Know the World: Alexandre Moreau de Jonnès and a Soldier's Devotion to Studying the Colonies (1818–1839)

Hugues Chabot, Pierre Crépel, Thierry Dumont, Christian Gaillard, Maryannick Lavigne-Louis, François Renaud, Denis Reynaud, Dominique Saint-Pierre, and Thomas C. Anderson

Abstract: Alexandre Moreau de Jonnès (1778–1870) is known above all for his military adventures throughout the world during his youth and, later, for his role as director of the ministry Statistique Génerale de la France (1833–1851). Before he became a public servant in the ministries after the Restauration, this character had also played the role of a *savant* and an intermediary to various scholars. His own contributions to knowledge, scattered in books and brochures, are concerned in particular with fauna, climate, and diseases of the colonies. Through his rich correspondence, this chapter examines not only what spurred the officer's acute curiosity, but also discusses the interaction and interdependence with established knowledge institutions and their members.

Keywords: France, military, correspondents, intermediary, Académie des Sciences

Why would a man, who was not a scientist by career or training, devote himself to scientific study? One might think of people of some level of privilege, such as aristocrats, doctors, abbots, lawyers, or the rich as those

This article, which was originally composed in a joint effort of the Commission Alexandre Moreau de Jonnès at the Académie des Sciences, Belles-Lettres et Arts in Lyon, has been translated from its original French by Thomas C. Anderson, who himself is a member of the same commission.

Bijleveld, Nikolaj, Arjen Dijkstra, and Samuel Gessner, eds. *Cultivating and Communicating Natural and Technical Knowledge around 1800: Devotees of Science.*
Taylor & Francis Group, 2025
DOI 10.5117/9789463728577_CH06

most likely to become involved in scientific knowledge accumulation in their spare time. One would also imagine that those with such privilege would typically be interested in providing armchair commentary on popular subjects like astronomy, botany, or archaeology. Alexandre Moreau de Jonnès (1778–1870) fits this mould well, and the fields he was interested in were those that reigned supreme at the beginning of the nineteenth century, when he conducted most of his fieldwork while living in the Antilles as a soldier. This figure, who later served as an administrator in many different offices over the course of several decades, had an unusual penchant for statistics, and imbued that interest into all of the other work he devoted himself to, including studies of geology, zoology, and geography. His varied career and enormous number of writings and correspondences make him an exemplary case study of an early nineteenth-century devotee of science who, as we will see, practised in the period just prior to the trend of professionalisation that would create barriers to entry for later figures to study science outside of established institutions.

This chapter presents an overview and analysis of scientific practice in the first half of the nineteenth century, a critical point just before the aforementioned trend towards professionalisation. As such, it does not embark on an exhaustive reading of Alexandre Moreau de Jonnès's writings, but instead proposes this singular figure as a case study and transitional figure in this broader historical point of inflection. An essential, and oft-overlooked aspect of the shift towards professional science in nineteenth-century France was the enduring presence and role of the many academies of science as sites of patronage and scientific community that fell somewhat outside the mandate of new Revolutionary-era centralised institutions. Therefore, this chapter endeavors to first provide this essential context of various political and institutional shifts affecting the practice of science in this period, before examining the importance and societal function of the work of devotees of science such as Moreau de Jonnès. The term 'devotee of science' is used here to replace the term 'amateur scientist' to avoid the pejorative sense of the term 'amateur', which would imply a certain level of incompetence or ineptitude, as 'devotees' did not lack technical competence, but were simply those who practised and studied science in an unpaid fashion. Fundamentally a man influenced by the contingencies of his time, Alexandre Moreau de Jonnès offers a rich example of one of the last great devotees of science, yet simultaneously helps punctuate this transition point due to the influence he had in his later career on these very same trends towards professionalisation. (Illustration 9)

Illustration 9. Portrait of Alexandre Moreau de Jonnès (1778–1870). Albumen print photo by Ch. Reutlinger (*Musée Carnavalet, Histoire de Paris, PH54227*)

Scientific Context of the Eighteenth Century

In both France and elsewhere throughout Europe, the pursuit of the ideals of the Enlightenment and the belief in the potential of science were all in full swing by the end of the seventeenth century. The trend to place ever increasing emphasis on rationality and empiricism that has come to be so emblematic of the Enlightenment steadily developed over the course of the eighteenth century. In France, the number of academies multiplied and scholarship in general flourished – the mathematical and physical sciences, natural history, archaeology, antiquities, and philosophy all grew in terms of membership and overall stature. In this period, it seemed as though the world and its many people, plants, and animals were knowable, and that the pursuit of universal knowledge was an attainable and worthy goal. To this end, the so-called encyclopaedic movement aimed to touch on every corner of Europe; but in particular, the *Encyclopédie, ou dictionnaire raisonné des sciences* (1751–1772) composed by Denis Diderot (1713–1784) and Jean le Rond d'Alembert (1717–1783), and its later editions – especially the *Encyclopédie méthodique* (1782–1832) – had a greater influence than other efforts, including in the eyes of the public.

Under the *Ancien régime*, the universities offered instruction in law, medicine, and theology, but contrary to contemporary developments in Germanic, Italian, and English-speaking countries, there were not any *facultés des sciences* nor schools for the humanities in Enlightenment-era France. Despite this, the sciences were far from absent from educational institutions. They were particularly important among the Jesuits and the Oratorians. The mathematical sciences were also taught in a number of military schools, which trained students in the three military sciences (*armes savantes*): artillery (mechanics, chemistry, and probability), engineering (geometry, architecture, and material studies), and maritime arts (fluid mechanics and astronomy). Beyond these, there were a diverse set of other institutions devoted to scientific research and knowledge acquisition – namely, botanical gardens, observatories, and veterinary schools, among others.

It is worth noting the importance of the Académie Royale des Sciences in Paris (hereafter referred to as the Académie des Sciences) as the principal location for research, exchanges between scholars, and as a mediator of interactions between other academies in both France and Europe more broadly. By the end of the eighteenth century, when Moreau de Jonnès was still in his youth, there were still very few scientific journals and periodicals besides the *Mémoires* of the Académie des Sciences and the series of annexes that this academy published on an intermittent basis. Outside of the Académie, the most important journal was the *Journal de physique*, organised by Abbé Rozier starting in 1772, which allowed both recognised and non-institutional scholars to share their work. Therefore, the Académie and its sister institutions in other French cities played a vital role in fostering scientific thought in a country that at that time had almost no other substantial infrastructure of science.

Throughout the eighteenth century, there were a number of people who, despite not being compensated, nonetheless dedicated themselves to the study of science within the framework of the academies. The disciplines which lent themselves most easily to this sort of 'amateur participation' were botany, mineralogy, antiquities, astronomy, and meteorology. By contrast, the mathematical and physical sciences, or those whose study necessitated the use of sophisticated instruments, invited a lesser degree of participation. In general, those who developed close connections with the academies were wealthy, and these same figures were sometimes named as 'corresponding members' or participated as candidates for the annual prizes sponsored by the academies. These prizes, or concours, began in the 1720s and became popular by the middle of the century. The competitions consisted of a specific academy publishing a precise question of some philosophical, intellectual, or practical importance, to which candidates had approximately a year to respond. Participation was open to everyone except titled members of the hosting academy, and candidates would enter by submitting complete, anonymised memoirs.[1] Once the deadline had passed, the academy would assign a commission to name (or sometimes refuse to name) one or multiple winners, or *lauréats*, and then would award the prize in a public ceremony.

Among the participants in the concours, we can cite two by way of example: François Etienne Dutour de Salvert (1711–1789) from Riom and Charles François Bicquilley (1738–1814) from Toul. The first was a nobleman with large landholdings, who owned farms and was a tax collector, both of which left him with long periods of leisure during which he could pursue other interests. For instance, he studied physics extensively, and focused on several other disciplines to a lesser degree. The second was a squire and *garde du corps du roi* (king's bodyguard), which were occupations that left him with considerably less free time. He primarily focused on mathematics and, in particular, he studied probability and its various applications. These two very different participants demonstrate how the prize money and esteem

1 This anonymity was at times more of an open secret, meaning that the true author was easy to identify, but was more authentic on other occasions.

associated with the concours compelled diverse groups of competitors to submit for the prizes.

Apart from the growth of what are today referred to as the 'hard sciences', the end of the century was also marked by the birth of the social sciences: men like Turgot, Condorcet, or the *Idéologues* serve as good examples of these developments. This broad-based scientific movement also impacted the French Revolution, in which many scholars such as Antoine Lavoisier (1743–1794) participated, particularly in the beginning and at the end. Several instances from this period, when old initiatives were extended or new ones were created, such as the founding of the École Normale in year III (1794–1795), the establishment of the École Polytechnique, or the Egyptian campaign, serve as good examples of scholarly impacts on the revolutionary period.

French Scientific Institutions after 1800

The evolution of scholarly institutions in the nineteenth century was a contradictory process. Undeniably, science and technology soared to new heights in this period and have only continued to grow since. At the same time, however, these developments spurred the anti-science movement dubbed the *réaction romantique*, or Romanticism. This was a broad-based movement, mainly taking place within the arts and humanities, in which scholars, authors, artists, and musicians all started to emphasise the value of the individual and the role of emotion in history and society writ large. Inspired by many of the ideals of the French Revolution and other contemporary debates happening throughout Europe concerning the limits of reason, Romanticism was a concerted reaction against the ideals of the Enlightenment. Therefore, in the early nineteenth century, rationalism and scepticism – the reigning modes of intellectualism in the preceding century – were increasingly out of vogue. However, while the anti-scientific views and caricatures put forth by François-René de Chateaubriand (1768–1848) in *Génie du christianisme* (The genius of Christianity) were celebrated and shared by many, a greater number of scholars chose to share Chateaubriand's general ideology while simultaneously practising and defending the sciences. This was the case for many important nineteenth-century scientists like the physicist André-Marie Ampère (1775–1836).

In 1808, Napoleon mandated the establishment of official faculties of science within French universities. While this would seem to be a positive reflection of science's growing importance and ascendency, from an institutional point of view, the creation of these faculties was mostly a formality. The primary role for professors was to conduct the *baccalauréat* exam for a handful of students each year while they taught very few courses and conducted minimal research.[2] What is more, these faculties were shut down at the beginning of the Restauration (the period from 1814 to 1830 of the Bourbon dynasty's restoration to the French monarchy, which was briefly interrupted in 1815 by Napoleon's one hundred days). Only the École Normale survived the initial shutdown, but it too was forced to close its doors in 1822. In this period, students such as Antoine Augustin Cournot (1801–1877) found themselves out on the streets and forced to pursue other ventures. One easily forgets that a new encyclopaedic movement developed at this very same time; Louis Gabriel Michaud's (1773–1858) *Biographie universelle* is the best known, but it was far from the only work of its kind in this period. In the same spirit, the 1820s saw an abundance of reviews concerning the scientific vocation. The best example would be the remarkable *Bulletin* compiled by the Baron de Férussac (1786–1836). This rich nobleman came from the south of France and was the son of a well-known officer and naturalist, and he himself was fascinated with natural history and in particular non-marine molluscs, both living and fossilised. Because of his wide-ranging interests and his statistical inclinations, he decided in 1822 to launch and direct the *Bulletin général et universel des annonces et des nouvelles scientifiques*, which was divided into eight subsections that were meant to cover all aspects of knowledge. The project was a general success up until its demise in 1831.[3] Alexandre Moreau de Jonnès published many articles in this periodical.

Faculties of the sciences and the humanities did not truly flourish until the July Monarchy (1830–1848), under François Guizot (1787–1874) and Abel-François Villemain (1790–1870). In Lyon, this launch would take place in 1835. This period also saw the establishment of the private scholarly associations, the *sociétés d'émulation* and their organisation within the Comité des travaux historiques et scientifiques (CTHS). At the same time, new scientific conventions established well-connected networks of participants and interlocutors, both before and especially after the advent of the railroads. This point marked a change in the role of the academies: before the 1830s, they constituted essentially the only place where scientists could

2 In France, the Baccalauréat is an exam taken at the end of high school, and is not equivalent to a Bachelor's degree (as it is commonly misunderstood to be).

3 More information on the *Bulletin* can be found in the presentation by Cournot, *Ecrits*, esp. pp. 15–31.

congregate, exchange research, and form community; after this point, the scientific life started slowly shifting towards the universities and the elite schools, as well as to public and private institutions devoted to engineering and administration. Little by little, the academies would come to serve for honorary *a posteriori* recognition of skills, career achievements, and for the distribution of awards.

The Re-establishment of the Academies

The academies, having been temporarily disbanded by the National Convention in August 1793, began re-establishing themselves in 1795, starting on the national level with the Institut de France. In Lyon, it took until 1800 for the prefect Raymond de Verninac Saint-Maur (1761–1822) to reconvene the academy and its members.[4] Outside of the academies, other intellectual societies also benefited from the era's renewed scholarly vigour and were able to re-engage with a range of intellectual projects on both the local and universal levels. Importantly, the academies continued their patronage of the concours prizes. Here, one other peculiarity of the academies must be mentioned: while they were not totally 'scholarly' institutions in the same sense (and without the same status) as the universities, they also were not state run, and therefore they occupied a sort of in-between status. As such, while the academies were forced to close during the so-called 'Terror' era of the Revolution, their members were recognised and offered patronage by both local and national authorities in other times.

This peculiar status was important when considering the process of naming members and identifying correspondents of the academies. By virtue of the freedom it was granted as neither a totally state-run institution nor a formal university, the Académie des Sciences (or the principal branch of the Institut de France), from its seemingly-imperial seat in Paris, was able to name associates and corresponding members from a wide range of social groups, including foreigners and non-Parisian French personalities, retirees, and others whose careers were not in the least associated with science. Therefore, the liminal space that the academies occupied between official state institutions and formalised universities overseen by the professoriate allowed their members a greater degree of flexibility in terms of what types of information they could consider, and from whom they might accept reports. It should be noted that the large number of so-called

4 See section 'Bureaucracy and Science in the French Restauration'.

enlightened devotees of science also reflected the state of contemporary scientific professionalisation since, through the first several decades of the nineteenth century, science remained a discipline that could be practised from pretty much anywhere, and by anyone.[5] By the 1830s, however, the onset of the Industrial Revolution began to slowly change these dynamics within the field. While this was by no means an immediate change that instantly expelled the devotees of science, or rendered their findings invalid, the increasing emphasis on ventures like mining, production of electricity, steam power, and other major areas of research into energy production and adjacent fields helped make professions like engineering and entrepreneurship more enticing. Gradually, as more and more money was invested in these areas, the impetus to become a professional scientist working for the government or a corporation grew, and with that growth, the role of the devotee of science slowly diminished. We flag this period as a major point of transition in the history of the professionalisation of science, and notably, Moreau de Jonnès conducted his fieldwork just prior to this shift.

The Military, Science, and the Young Alexandre Moreau de Jonnès

Having established the state of both the academies and science more broadly through the 1830s, it is now time to focus more specifically on Alexandre Moreau de Jonnès (1778–1870). Per Parisian birth records, he was baptised the day after he was born on 20 March 1778 in the Saint-Paul church in Paris, though his father was conspicuously absent. His mother, Charlotte Nogué, was recorded as living on the Rue Vieille du Temple, in the Marais quarter, while his father, Silvestre Jonnès, a bourgeois of Paris, was a fugitive.[6] Alexandre was their illegitimate child.[7] In his memoirs, however, which were posthumously published in 1890 by Alphonse Aulard, Alexandre wrote of being born on 19 March, 'or, more likely, a fortnight

5 Though, it should be noted that numerous works in the history of science and science and technology studies (STS) over the last thirty years have convincingly argued that, while European scientific institutions remained open to participation to a wide range of *European* actors, these institutions also routinely excluded non-white and female actors. Despite their exclusion from the canonical narrative, these groups certainly practised science regularly, and future studies would do well to remember and more seriously consider their contributions.

6 Anon., 'Naissances', fols. 43–44. 'fille majeure demeurant susdite rue Vieille du Temple'.

7 Moreau de Jonnès, *Mémoire*, p. 352.

earlier' on the banks of the Rance, in the outskirts of Saint-Malo.[8] Given his propensity to embellish his stories, it is difficult to trust this autobiographical origin story over the official birth records. What is totally clear is that he consistently emphasised his Breton heritage (his mother's lineage) throughout his life, and that he devised a false aristocratic name for himself at a young age in order to claim relation to a number of famous Moreaus who were, in his account, 'a Gaelic clan who produced many remarkable and [strong] sons'.[9]

Despite this somewhat fraught infancy, Alexandre wrote of passing a 'golden childhood' in Paris, living in the Marais quarter under the protection of the Robien family from Rennes.[10] In 1786, at the age of eight, he began school in Rennes and later praised the school's pedagogy as being far ahead of its time. In February 1791, when he was just thirteen, he was obliged to return to his childhood home in the Marais due to the state of the Revolution. Reflecting on this time, he wrote, 'my tall stature, my small black moustache, and my having the overall appearance of a Breton federate condemned me to being conscripted for service by the Minimes [National Guard], who selected me to serve in their battalion'.[11] After witnessing the early days of the Revolution while simultaneously devoting himself to study at the local library under the care of the librarian Ameilhon, Alexandre returned to Rennes in September 1792 to continue his studies. Unfortunately, little information remains as to the details of his studies, such as the types of books he read. What seems clear, however, is that the circumstances of the Revolution drove him to become an autodidact, and that it was not by pure choice that he became his own teacher.[12]

Just weeks after Alexandre returned to school, on 10 October 1792, the National Convention declared a state of emergency, claiming that France was in danger. In response, the mayor of Rennes made a call-to-arms to the students of the city, and the next day they all signed their commitment papers. Suddenly, Alexandre found himself in the Morbihan region of Brittany fighting against the counter-revolutionary Chouannerie uprising

8 Moreau de Jonnès, *Mémoire*, pp. 352–67, 448–66, 550–65.

9 Moreau de Jonnès, *Mémoire*, p. 354.

10 Moreau de Jonnès, *Mémoire*, p. 454.

11 Moreau de Jonnès, *Aventures*, vol. 1, p. 460.

12 More generally, Moreau de Jonnès's personal papers and correspondences seem either lost or destroyed. Certainly, some of his letters can be found scattered throughout the archives of his correspondents, but an exhaustive index of these tracings was beyond the scope and capabilities of the present study.

(1794–1815). He became a navy gunman (*artilleur de la marine*) in Brest, and later was named the Master-at-Arms (*capitaine d'armes*) aboard the brig *Papillon*: 'Soon thereafter they put me in charge of accounting for all the ship's materials [...] it was the first official statistical charge I received; I was sixteen'.[13] From that point until his resignation from the military in Fort-de-France, Martinique in 1815 and his subsequent return to Paris with his young family, Alexandre embarked on a series of incredible adventures (that were reasonably well embellished in his *Aventures*), including voyages to the Antilles, a stint in a London prison, and so on. During this time, he also enjoyed a *beau mariage* in Fort-de-France in 1806 which helped ensure his source of revenue. He also studied copious legal texts of his own accord and out of self-driven interest.[14] It was from observations that he made during this period of adventure that Alexandre would base the majority of his future scientific work.

Through these events, we see that young Moreau de Jonnès the soldier was in truth more of a scholar-soldier. He certainly participated in the so-called *armes savantes* of the day, or the tendency for educated men to take up extra-military interests during their years of service. What is especially striking is his level of curiosity from a young age, especially given that he was almost entirely self-taught. He was fairly well acquainted with members of high society such as Beaumarchais (1732–1799) and the Robien family; yet on the other hand, he was exposed to a number of extraordinary and violent events during the Revolution. His testimonies reveal that he quickly matured beyond his age and that his superiors came to entrust him with tasks typically reserved for older or more experienced people. As just one example, while on his various military campaigns, he was the designated intellectual of the regiment who was in charge of all the accounting and general management. In short, the books his battalion carried were for him, and he would carry his love for statistics and penchant for writing with him long after his military days had ended.

13 Moreau de Jonnès, *Aventures*, vol. 1, p. 147. 'Bientôt on me chargea de toute la comptabilité du matériel [...] C'est la première statistique officielle qui m'ait été confiée; j'avais 16 ans'.

14 Throughout this biography, we refer to the work currently being prepared for publication by the Academy in Lyon, where one will find a complete biography of Moreau de Jonnès. For the period prior to 1815, the principal sources are his own diaries and journals (often to be taken with a grain of salt, since they are embellished) and two more-or-less autobiographical novels. But, it must be noted that there remain numerous uncertainties in Alexandre's biography and that not every detail is able to be verified. We have done our best as a collective research team to recount only those details which can be corroborated or that seem irrefutably true, but parts or all of this account may nonetheless contain errors.

Becoming a Correspondent of the Académie des Sciences circa 1816

What did it mean to be a correspondent of the Académie des Sciences in the French Restauration, and how would one achieve this title? Was this something one would do for pleasure or personal satisfaction, or was this a stepping stone in one's career? In almost all cases in this period, correspondents were people who worked in scientific professions in a broad sense – it should be noted that they were exclusively men. Correspondents could be teachers, astronomers, directors of botanical gardens, curators of natural history cabinets, physicians, surgeons, veterinarians, pharmacists and apothecaries, architects, engineers, or agronomists, and most did not live in the Parisian metropole. Primarily, the value of this position was in receiving recognition, integrating oneself within scientific networks, and attaining better access to new scholarly developments and information. The obligations for a corresponding member were (and still are) fairly light, and essentially consisted of updating the academy from time to time about one's work or that of one's close contacts.[15]

The first mention of Moreau de Jonnès in the minutes of the Académie des Sciences regards the receipt of samples of volcanic ash on 26 October 1812. Next, in 1814, and while he was still serving as an aide de camp to the general Houdetot, Alexandre sent several memoirs on the volcanoes of the Antilles. These two initial mentions aside, it was not until 1816, once he had returned to Paris and was no longer serving in the military, that Moreau de Jonnès truly engaged in a regular and uninterrupted correspondence with the Académie. In that year, on 5 and 12 August, he read his *Monographie du Trigonocéphale des Antilles*, an account of the lancehead viper, to the institution's members. The Académie des Sciences dispatched a commission of three members (Lacepède, Geoffroy Saint-Hilaire, and Duméril) to examine and scrutinise it. One month later, on 9 September, André Marie Constant Duméril (1774–1860) delivered their glowing review of Alexandre's work, stating that the poisonous snake had been very well-studied and with 'a new level of precision', even if the memoir's author was 'not quite familiar with the terms used by anatomists'.[16] In truth, this memoir probably appealed to the Académie des Sciences not only for its scientific quality, but also due to the relative dearth of reliable scientific information on the Antilles that was available at the time. Alexandre was immediately 'placed

15 For an example of these standards in 2022: Institut de France, Académie des Sciences, *Statuts et règlement intérieur*, https://www.academie-sciences.fr/pdf/admin/statuts.pdf.

16 Institut de France, Académie des sciences, *Procès verbaux*, pp. 77–78.

on the list of people with aspirations to become corresponding members', and was rapidly admitted to the position on 25 November 'in the section on geography and navigation'.[17] Three years later, he would also be named the first laureate of the Montyon prize in statistics by the same academy.

The first thirteen publications that Moreau de Jonnès wrote all concerned the Antilles. After 1816, the Académie's minutes show a veritable outpouring of submissions (practically one a month beginning in 1816) concerning a variety of topics including geology, geography, plants, animals, humans, health, and several other subjects.[18] Evidently, Alexandre was in regular contact with the full members of the Académie, most especially its three rotating secretaries, Delambre, Fourier, and Cuvier. This correspondence is seen in the many letters sent to the Academy in Lyon (l'Académie des sciences, belles-lettres et arts de Lyon, or ASBLA). What's more, he made several attempts to be elected a fully titled member, though he failed three separate times on 30 July 1821, 9 January 1826, and 25 January 1830. We do not know how he reacted to these failures or if they deterred him from further submissions. In any case, Moreau de Jonnès clearly enjoyed participating in the concours and corresponding with the different academies, as he made clear in the frontispieces of his books. It is difficult, however, to determine whether Moreau de Jonnès submitted his work to make contacts or if he was unconcerned with making connections and was just a true devotee of science. Given the conditions of his childhood and constantly upwards career trajectory, however, it was almost certainly a mix of the two.

Bureaucracy and Science in the French Restauration

Did bureaucrats and administrators receive a baseline of scientific training during the Restauration? For the most part, no, apart from a bit of medical training. For example, the administrative offices listed in the *Almanach de Lyon* are 'man of the law' (*homme de loi*), judge, landlord, banker, merchant, military officer, physician, or pharmacist – rarely were there exceptions to this list. In the majority of cases, therefore, bureaucrats studied law. Consequently, in this period there were neither official, governmental scientific institutions nor people with scientific training influencing politics

17 Institut de France, Académie des sciences, *Procès verbaux*, pp. 78, 113.

18 See, among others: Alexandre Moreau de Jonnès, *Essai sur l'hygiène militaire des Antilles*, (Paris, 1817).

in France. This administrative shortcoming extended beyond what would today be termed the 'hard sciences', as well. Even though the Académie des Sciences Morales et Politiques was added to the Insitut de France in 1795 to support the study of the social sciences, Napoleon disbanded the fledgling institution in 1803 since he found the academy and its members to be too intractable. Therefore, between 1803 and 1832, there was no official place for sociological or anthropological work within the formal French institutional hierarchy, and those seeking to study those fields had to search for other avenues to meet and share their work.[19] This was the fraught state of science, broadly construed, during the time when Moreau de Jonnès was conducting the majority of his scientific work.

Considering that Moreau de Jonnès developed his penchant for numbers, accounting, and statistics at a young age and that he incorporated these into both his scientific and political work, it is necessary here to provide more specific context on the state of the field of statistics between 1800 and 1830.[20] In brief, in the eighteenth century there were three main statistical traditions: German statistics attempted to describe and tally a country's productive forces; English statistics were more conjectural, largely based on tables of mortality and on insurance; and French statistics focused on political arithmetic, culminating in Condorcet's social mathematics and the foundational works of social science. This system of French statistics is what is of concern here. Starting during the Directoire (1795–1799), François de Neufchateau (1750–1828) launched an initiative to compile a comprehensive set of data about France's *départements* with the help of the academies. This project continued during the years of the Consulate (1799–1804), during which time an official statistics of the prefects (*statistique des préfets*) was established by the memorandum of Chaptal on the 19 of Germinal, year IX (9 April 1801). Under this system, the prefects, or administrators of each of France's *départements*, wrote to the leaders of the various learned societies in a region, delegating them the task of compiling a complete description of their institution and what they knew about the surrounding region. Several remarkable reports resulted from this system, though many others were compiled in a hurry and were quite hasty with many oversights. In the following decades after this initial period, these statistical systems experienced highs and lows depending both on changing policies and, at

19 See, for example, Vincent, 'Les "sciences morales"'.

20 There are a number of historical works that cover these developments better than we can here. For a list of the specific context of Lyon, see: Crépel, 'Les statistiques', where a list of the most important historical works can also be found.

times, on who was seeking to promote them or not (ministers, prefects, scholars, physicians, etc.). Therefore, while statistics fared better at times during the Restauration than the other social sciences, there were also long periods in which the type of statistical analysis that Moreau de Jonnès was passionate about would have been markedly out of vogue.

Thus, while living in Paris in the 1820s, Moreau de Jonnès worked for a diverse set of governmental offices such as the naval and colonial office and the ministry of commerce. His position varied depending on the tenor of the moment. For example, he held an important position within the moderate Martignac government (1828–1829), during which time he worked alongside the Minister of Commerce Pierre de Saint-Cricq (1772–1854) and was able to devote his time to compiling data on a variety of topics including economics, demographics, and production statistics. At other points he held less-significant positions. Due to the volatile nature of his employment and of the field more generally, it was not until several years into the July Monarchy that Alexandre Moreau de Jonnès became the leading figure in French statistics, starting in 1833 or 1834. As he explained in his *Elémens de statistique* in 1847, he was an ardent supporter of exhaustive censuses, and was an equally steadfast opponent of surveys, questionnaires, estimations, and averages. Over the next several decades, he led several nation-wide census projects and helped influence the future course of French statistics.

The Career and Works of Moreau de Jonnès

It would be a much bigger project than we are able to conduct here to give a true overview and analysis of the complete works of Alexandre Moreau de Jonnès, especially given the wide range of scientific disciplines he contributed to. One might say that, due to the high volume and wide range of his work, writing was somewhat of an obsession for him.[21] In terms of his engagement with the natural sciences, all reports from the Académie des Sciences in Paris indicate that his communications were received with interest – there is little room for doubt about that. At times, members openly gave their praise – for example, on 26 December 1825, Portal, Pelletan, and Boyer concluded their report on Moreau de Jonnès's description of acute cholera by writing: 'this should serve as a model for anyone proposing to

21 Therefore, the following remarks should not be misinterpreted as an exhaustive representation of his entire curriculum vitae, but rather as pointed remarks about several areas of specific interest.

enrich this essential branch of science'.[22] Although being that of just a devotee of science, his work was clearly valued and taken seriously by the same academy that repeatedly denied him full member status for years.

What was the nature of Jonnès's scientific state of mind? Often, he proceeded in an exhaustive fashion, enumerating and measuring different phenomena, species, and different products he encountered seemingly first-hand.[23] He sought to work with long, complete datasets, both in terms of time and space. He was hostile towards probability and estimation, preferring precision and quantification wherever possible. His work is full of careful, assiduous descriptions. But his analyses also went beyond describing for description's sake and counting for counting's sake – he aimed to comprehend things, to investigate causality, and to understand the linkages between different phenomena. Consider for instance one of his typical phrases: 'Understanding the Antillean climate is more than just an achievement for the physical history of the globe, it is also a great acquisition for the medical and administrative sciences'.[24] Another example of this deep interest in comprehending complex phenomena and relationships is represented in his writings on the spread of yellow fever and cholera. Here, he drew explicit links between geography, calculus, and social utility. This work was one of his key contributions, yet it still reveals how statistics were the true crux of everything he worked on. Moreau de Jonnès actively worked to accumulate his own data at every opportunity; from time to time, he cited official reports, but even then, it is difficult to ascertain whether the numbers he cited actually came from the reports or from his investigations and fieldwork. A tally of the works he conducted while directing the French *Statistique Générale* would without doubt enable a better understanding of his research methods.[25]

22 'il mérite de servir de modèle à ceux qui se proposent d'enrichir cette branche essentielle de la science'.

23 It should be noted that often, in the colonial context in particular, publishing authors claimed firsthand experience that was at times not completely genuine. While there is no direct evidence of it here, it is not unlikely that Moreau de Jonnès relied on the expertise of subjugated interlocutors like enslaved Africans, various indigenous peoples, and women in the formation of his different knowledge claims from the Antilles.

24 Moreau de Jonnès, *Histoire*, p. 159. 'La connaissance du climat des Antilles, n'est pas seulement une acquisition pour l'histoire physique du globe, c'en est une pour les sciences médicales et administratives'.

25 This would, however, demand an enormous amount of archival work in the F12 (*commerce*) and F20 (*statistiques*) collections that we have not yet been able to conduct. Furthermore, he made several drawings (both schematics and landscapes), of which several survive. These can be found in the Archives Nationales d'Outre-Mer (ANOM in Aix-en-Provence), in the Auxerre library, and in the British Library. An overview will be provided in the forthcoming volume being developed by the Academy in Lyon (ASBLA).

One of the constant themes seen in Moreau de Jonnès's scientific work, which was certainly a broader contemporary trend shared amongst his peers, was his constant concern over the budding popularisation of science. This concern manifested itself in a veritable frenzy of publications, not solely by traditional press houses and in academic reviews (in which he published more than sixty books or opuscules between 1816 and 1873), but also in a multitude of other articles, manuals, and reports. To date, we have found publications in at least a dozen periodicals: the *Revue encyclopédique* (more than twenty contributions between 1819 and 1833), the *Annales générales des sciences physiques*, the *Annales maritimes et coloniales*, the *Annales politiques*, *Le Courrier français*, *Le Globe*, the *Journal de physique, de chimie et d'histoire naturelle*, the *Journal des voyages*, the *Journal du commerce*, the *Revue de Paris*, and the *Revue des Deux Mondes*.[26] Did Moreau de Jonnès care more about receiving official recognition from the academies, publishing for a broad audience through these various outlets, or creating administrative reports for the government? This question remains a mystery, and the author himself left no indication of his sentiments in his copious writings (most of his personal papers, however, are lost). One possibility is that Moreau de Jonnès viewed these three different manners of professional expression as complimentary categories.

Comparisons with Several Contemporaries

One could make any number of comparisons between Moreau de Jonnès and well-known scholars of his time, professors in faculties of science, or members (titled or correspondents) of the Académie des Sciences. Here we present four contemporaries, who share similarities with Alexandre in terms of their titles, the types of work they conducted, and more importantly, in their status as devotees of science and writers. Of these four figures, two – Réveroni Saint-Cyr and Mathieu Bonafous – were correspondents of the Academy of Lyon (ASBLA), one – Ozanam père – was declared the laureate of three different concours from the Académie des Sciences, and one – Jean-Baptiste Ricord – was a physician who was interested in natural science, conducted work in the Antilles at around the same time, and similarly to Moreau de Jonnès, sought metropolitan recognition for his colonial field work.

26 There are likely more, and this list does not represent an exhaustive effort to compile his works. It is meant as a representative sampling.

Révéroni Saint-Cyr (1767–1829)

The career of Jacques Antoine Révéroni Saint-Cyr provides many easy comparisons with Moreau de Jonnès: born ten years apart, both men staked dubious claims to nobility, retired from the military within a year of each other, and both found a sort of legitimacy in their recognition as corresponding members of the Academy of Lyon. Contrary to Moreau de Jonnès, Révéroni enjoyed a traditional education and was trained in mathematics, mechanics, and physics at the École royale du génie militaire in Mézières. From a young age he aspired to literary glory, publishing eight novels and twenty-one comedies, whereas Moreau de Jonnès took over sixty years to publish two novels, since his interests lay elsewhere. The two men, however, shared a confidence in the ability of science to solve humanity's problems. On the 29 of Ventôse in the year XII (20 March 1804), Marc-Antoine Petit read his report on Révéroni's *Essai sur le perfectionnement des beaux arts par les sciences exactes* (Essay on the perfection of the fine arts by way of the exact sciences) which the latter had recently sent to the Académie des Sciences: 'The hour is come when every artist must also be a scholar […] Applying physics and calculus to all the arts, to music, to painting, Mr Révéroni demonstrates the means, more or less extraordinary, yet always wise, to perfect them'.[27] Throughout his latter years, Révéroni attempted to make all aspects of human conduct conform to mathematical principles, and he sank into a sort of madness of over-rationalisation, which Moreau seems somehow to have avoided.

Jean Antoine Ozanam (1773–1837)

Son of the royal notary, clerk for the notary in Point d'Ain, soldier in the year II (1794), and later officer who returned to civilian life in 1800 to work as a silk merchant, Jean Antoine Ozanam already had a respectable career before he went to Italy to become a physician in 1810, and especially by the time he returned to Lyon in 1816 to take up a position at the Hôtel-Dieu. He was the father of Frédéric Ozanam (1813–1853), who would go on to establish the church's social doctrine. For his part, doctor Ozanam is cited three separate times in the concours of the Academy: in 1825, he received a prize medal for his method for removing sericin from silk; in 1829, he received an honourable mention for his *Essai historique et statistique* (Essay on history and statistics) covering, in particular, religion in Lyon; and in 1832, his *Essai*

27 'L'heure est venu où tous les artistes doivent être des savants […]. Appliquant la physique et le calcul à tous les arts, à la musique, à la peinture, Mr Reveroni indique des moyens, plus ou moins extraordinaires, mais toujours judicieux, pour les perfectionner'.

de statistique du département du Rhône (Statistical essay on the Department of the Rhône) won the grand prize.[28] He also participated in the concours on the eulogy of Jacquard in 1837, but did so without success. Therefore, in the career of this former soldier, traveller, and doctor, one notes an interest in various different scientific and technical questions. His *Histoire médicale générale et particulière des maladies épidémiques contagieuses et épizootiques qui ont régné en Europe depuis les temps les plus reculés jusqu'à nos jours* (General and particular medical history of contagious human and animal diseases from the earliest times to the present day, in four volumes), which he self-published in 1835, represents a study that was at once intimately linked to his profession yet also fundamentally utilitarian.

Mathieu Bonafous (1793–1852)

Mathieu Bonafous was born in Lyon and died in Paris, but he spent the majority of his years living in Turin. This agronomist published a number of works on silk, maize, and other agricultural products. He also served as director of the botanical gardens in Turin. He was elected as a corresponding member of the Academy in Lyon (ASBLA) in December of 1825 (and later named an associate in December of 1837), and he submitted memoirs on the mulberry tree and the silkworm in 1830. From that point, he continued to submit a number of works on a variety of topics until 1847. He funded three different prizes, one for the eulogy of the Abbé Rozier between 1830 and 1832, another for the critical translation of Virgil's Georgics from 1835 to 1837, and the last for the history of silk in 1839. He participated in a number of other foundations, most notably in Savoy. He was elected a corresponding member to the national Académie des Sciences in the section on rural economies in 1835. Therefore, while he clearly had a lot of success in the concours, one sees that this agronomist dedicated himself to sciences related to his profession and did not devote the same level of attention to larger projects across diverse fields as did Moreau de Jonnès.

Jean-Baptiste Ricord (1777–1837)

Jean-Baptiste Ricord was a colonial physician in Guadeloupe working around the end of slavery in the French Empire. He was born in Martinique and completed his medical studies in New York before returning to the Antilles. Initially, Ricord worked in a hospital in Martinique, most likely in

28 These memoirs are preserved in the archive of the academy in Ms 247, 252, and 286, respectively. An account of the presentation of these different awards can be found in: Dumas, *Histoire*, pp. 510–48.

Fort-de-France, before moving to Guadeloupe's Grande Terre to pursue his own private medical practice and research. In Guadeloupe, he conducted a series of experiments with poisons and toxic fish indigenous to the Antilles, most notably examining the manchineel (*Hippomane mancinella*) and nhandiroba (*Fevillea cordifolia*) plants, as well as a certain fish that he named 'that which the slaves [sic.] call *never dead*'.[29] Ricord attempted to become a corresponding member of the Académie Nationale de la Médecine by presenting a scientific and medicinal treatise, conceived as a survey of the state of the colonies in the waning days of slavery, which he wrote between 1820 and 1826, roughly the same period when Moreau de Jonnès presented the majority of his research about the Antilles. He engaged in a correspondence with the Count of Lardenoy, then-Governor of Guadeloupe, in order to gain the support of both the count and the minister of the naval and colonial administration. In this letter, he explained that his work was of the utmost importance to the colonial project:

> The importance of [my book's] subject, which concerns itself with the life of the colonists and that of all the beings who live in their midst, cannot possibly fail to seize your attention, you who are loved like a father in these parts, and from whom we have the right to claim in return all the tender care of which you are capable of giving. I dare hope, Monsieur le Comte, that you will take interest and share this with his excellency the minister of the colonies for, in removing the veil beneath which hide the poisons of the Antilles, we will unmask the wicked ones [who seek to use them] and which, by instructing men about them without distrust, will provide them with the means to defend themselves from the enemies who surround them.[30]

These two men, Ricord and Moreau de Jonnès, devoted themselves to two different branches of a single unified project. One, a little-known colonial physician, experimented with poisons in order to defuse the perceived threat of the archetypical 'slave poisoner', while the other created an

29 Ricord, 'Mémoire', p. 26. 'Un poisson que les nègres appellent jamais mouri'.

30 Ricord, 'Lettre', p. 2. 'L'importance de son sujet, qui n'intéresse rien moins que la vie du colon et celle de tous les êtres qui l'environnent, ne peut manquer [de] fixer sérieusement votre attention, vous qui êtes ici aimé comme un père, et dont on a le droit de réclamer en retour, toute la tendre sollicitude dont vous êtes susceptible. J'ose espérer, Monsieur le Comte, que vous vous intéresserez au près de son Excellence le Ministre des Colonies, pour celui qui, en écartant le voile qui cache les poisons des Antilles, démasquera les méchants, et qui, en instruisant l'homme sans défiance, lui fournira les moyens de se garantir des ennemis qui l'environnent'.

exhaustive list of exploitable resources, people, and economic potential in France's colonial territories. Both figures were involved in a reassessment of the values of the colonial system during a time of colonial instability in the Antilles (just fifteen to twenty years after the Haitian Declaration of Independence and two decades before the definitive abolition of slavery throughout the French Empire). Neither was a professional scientist or titled member of any academy like Hans Sloane (1660–1753), to name one notable English corollary from the previous century, but their collective works nonetheless demonstrate an important point about the state of the professionalisation of science by the 1830s in France. Namely, for economic and social reasons that were intimately tied to the decline of the chattel slave system, in this early nineteenth-century period, there was space for devotees of science to engage in new types of investigations and experiments that would not have been conceivable and also would not have been seen as necessary mere decades earlier. Theirs was an extremely politicised science that was formulated with the specific goal of justifying the French system of empire in a colonial theatre that was coming to be seen as increasingly volatile, decreasingly profitable, and ever-more untamed.

Conclusion

It would be a mistake to take these few examples and draw sweeping conclusions about those who devoted themselves to scientific study without being compensated financially. There are, however, several trends that have emerged in our examinations of these different historical snippets of science, statistics, and colonialism more broadly. First, the majority of scientific correspondents who engaged with the academies on either a national or a local level seem to have been engaged in a profession that was at least vaguely related to science or technology. The several rich landowners and ecclesiastics amongst their ranks do not truly contradict this trend. Therefore, these devotees of science tended to occupy themselves with questions and subjects that were fundamentally similar to or intertwined with their primary work. These projects generally enriched their knowledge base and expanded their horizons in an immediately useful way, though at times these devotees of science launched into investigations purely out of some natural affinity or internal desire.

This does not appear to have been the case for Moreau de Jonnès, however: his various roles in the military (during which he mostly studied geography) and France's administration (when he was in charge of economic matters) did not prevent him from developing a penchant as well for the natural sciences, even including their non-utilitarian aspects. In short, he had interests that extended beyond his professional occupation and for him, the choice of possible materials and research questions was always fairly broad. For example, while in the Antilles, Alexandre did not hesitate to express on a number of occasions that he took advantage of his circumstances to indulge and investigate his full range of passions and interests in the natural sciences.

Moreau de Jonnès, however, was more the exception than the norm. Typically, devotees of science held a few personal interests that they pursued for a combination of personal or professional gain and genuine curiosity. If, in fact, there ever was a bias towards certain subjects or materials amongst the majority of those who did not work in para-scientific jobs or did not have 'one foot in, one foot out' in terms of scholarly activities, then that bias lay in whatever may be the subject presented by a given concours. They frequently competed for the prizes, likely in pursuit of their monied awards, though it should be said that they were seldom selected as the winners by the titled members of the academies. With all of this considered, we are left with the simple question: could devotees of science ever truly pursue a pure passion project, or were their investigations always bound by the same types of questions of prestige and compensation as their professional counterparts?

How, then, should one compare Moreau de Jonnès to other intellectuals, and in particular to the titled members of the Académie des Sciences, who were each versed in disciplines that were similar to Moreau de Jonnès areas of interest like geology, zoology, epidemic disease, geography, and statistics? One might respond to this question in three points. First, he was a scholar of considerable esteem in his own right, who was respected by many. Among those with whom Moreau de Jonnès fostered a close relationship were two of the perpetual secretaries of the Académie des Sciences, Georges Cuvier and Joseph Fourier, who he referred to on several occasions as 'my illustrious friend'. Second, even though his areas of interest show a coherence around questions concerning the Antilles and the circulation of living creatures around the globe, as a scholar he was far less specialised than those who gained more acclaim in their own right for their intellectual work. Finally, Moreau de Jonnès was not paid for his scientific work, but gained his money elsewhere via a different, rather time-consuming profession as an administrator in various different ministerial offices. The main function of this work was therefore not on science, but was instead on the statistics of foreign trade. These conflicting interests may at last help to explain why, for example, he remained a corresponding member rather than being named a titled one by the Academy of Lyon. Furthermore, it may offer an explanation as to why he would be elected (late, it may be added) to the Académie des Sciences Morales et Politiques, despite his enduring inclination towards the natural sciences. Alexandre Moreau de Jonnès was therefore a true devotee of science in that, even though his profession consistently seemed to divert his attention or pull him away in various other directions, he remained steadfastly interested in science and continued to foster his own interest in various fields of scientific inquiry throughout his life.

Bibliography

Anon., 'Naissances', 5Mil 54, fols. 43–44. Actes de l'état civil reconstitué 1872. Archives de Paris.

Cournot, A.A., *Ecrits de jeunesse et pièces diverses*, in: B. Bru and T. Martin, ed., *Oeuvres completes*, vol. 11, parts 1–2 (Paris 2010).

Crépel, P., 'Les statistiques et l'Académie de Lyon dans la première moitié du XIXe siècle', in: *Mémoires de l'Académie des Sciences, Belles-Lettres et Arts de Lyon* ser. 4, vol. 21 (2021), pp. 181–98.

Dumas, J.-B., *Histoire de l'Académie Royale des Sciences, Belles-Lettres et Arts de Lyon*, vol. 2 (Lyon 1839), pp. 510–48.

Institut de France, Académie des Sciences, *Procès verbaux des séances de l'Académie*, vol. 6: *1816–1819* (Hendaye 1915).

Institut de France, Académie des Sciences, *Statuts et règlement intérieur: Actuels et anciens* (Paris 2022). www.academie-sciences.fr/pdf/admin/statuts.pdf [accessed 29 April 2024].

Moreau de Jonnès, A., *Monographie du Trigonocéphale des Antilles* (Paris 1816).

Moreau de Jonnès, A., *Essai sur l'hygiène militaire des Antilles* (Paris 1817).

Moreau de Jonnès, A., *Histoire physique des Antilles françaises*, vol. 1 (Paris 1822).

Moreau de Jonnès, A., *Aventures de guerre au temps de la République et du Consulat*, vols. 1–2 (Paris 1858).

Moreau de Jonnès, A., *Mémoire sur les colonies et lettres à l'Académie de Lyon* (Critical and annotated edn; Lyon 2023).

Ricord, J.B., 'Mémoire sur le mancenillier vénéneux, faisant partie d'un ouvrage sur les plantes vénéneuses, les insects et les poisons vénéneux, les vers, les méduses, les physalides, et les crustacés qui se trouve à la Guadeloupe', 1822. Beinecke Lesser Antilles Collection, Hamilton College, Clinton, NY, USA.

Ricord, J.B., 'Lettre à son excellence, Monsieur le Comte de Lardenoy, Lnt. Général, Gouverneur et Admtr pour le Roi dans l'Isle de la Guadeloupe', 6 January 1823. Beinecke Lesser Antilles Collection, Hamilton College, Clinton, NY, USA.
Vincent, J., 'Les "sciences morales": De la gloire à l'oubli?: Savoirs et politique en Europe au XIX siècle', *La Revue pour l'histoire du CNRS* 18 (2007). https://halshs.archives-ouvertes.fr/halshs-00453627 [accessed 29 April 2024].

About the Authors

Hugues Chabot is a senior lecturer in history and philosophy of science at University of Lyon 1, and corresponding member of the Académie des Sciences, Belles Lettres et Arts de Lyon.

Pierre Crépel is a historian of mathematics, former research fellow of the CNRS, and lecturer at the University of Lyon 1.

Thierry Dumont is a former research engineer at Institut Camille Jordan, Lyon.

Christian Gaillard is emeritus professor of geology at University of Lyon 1.

Maryannick Lavigne-Louis is the former general-secretary of the Society Préinventaire des monuments et richesses artistiques du Rhône, Lyon.

François Renaud is emeritus professor of microbiology at University of Lyon 1.

Denis Reynaud is emeritus professor of language and French literature at University of Lyon 2.

Dominique Saint-Pierre is a lawyer, former deputé of the French National Assembly, and member of the European Parliament.

Thomas C. Anderson is a Ph.D. candidate in the joint programme in History of Science and Medicine and Early Modern Studies at Yale University.

7. Putting Knowledge to Use: Clergy and Science in Denmark and the Netherlands around 1800

Nikolaj Bijleveld

Abstract: This chapter studies the role of Dutch and Danish pastors in the advancement and dissemination of natural and agricultural knowledge around 1800. The comparison between these members of the Danish Lutheran state church and of the Dutch Protestant churches illustrates that their activities were typical neither of nation nor of denomination, and did not necessarily depend on the constitutional position of the church. It argues that the position of these pastors as knowledge brokers relates to the commonalities in their social and cultural context obtained through their vocation. They possessed the necessary skills and tools, had access to the right networks, and could be acknowledged as protagonists of international, moderate Enlightenment ideals of obtaining national prosperity through putting knowledge to use.

Keywords: pastors, religion, knowledge brokers, agriculture, societies

In 1771 the Danish pastor Johan Arndt Dyssel (1726–1795) presented a clear statement on the close connection between clergy and science:

> the sciences that are of real use to a people, those that shall help the country in hard times, those that will free the people from poverty, and the realm from ruin, these sciences should be spread, they should be

Part of the research on the pastors presented here was conducted as part of my PhD trajectory, which resulted in Bijleveld, *Voor God, volk en vaderland*, Bijleveld, 'Language, National Culture and the Clergy; and Bijleveld, 'The Nationalization of Christianity'.

Bijleveld, Nikolaj, Arjen Dijkstra, and Samuel Gessner, eds. *Cultivating and Communicating Natural and Technical Knowledge around 1800: Devotees of Science.*
Taylor & Francis Group, 2025
DOI 10.5117/9789463728577_CH07

learned, practised and propagated, next to the knowledge of God in every corner of the country; And how is this better done than when the clergy, who live in the whole circumference of the realm, are given material to work both practice and theory.[1]

Many pastors were convinced of the fruitful dynamics between church and science. Among the devotees of science, we find theologians who became renowned in academia as mathematicians or agriculturalists, as well as pastors who were prominent members of learned societies and worked on natural philosophical and agricultural experiments from their rectories. Oftentimes they did so in reference to religious as well as to educational and patriotic ideals.

This chapter examines the ideas and actual participation of the Dutch Reformed and Danish Lutheran clergy in the advancement and dissemination of knowledge related to natural philosophy and agriculture in the decades around 1800. Studying both groups in both countries improves our view of the activities that in general were undertaken. A comparison between the protestant clergy in the Netherlands and Denmark is furthermore instrumental in obtaining a better understanding of especially the ecclesiastical, social, and cultural context and therefore the factors that were of importance for these activities.

First, this chapter introduces the differences and similarities between Denmark and the Netherlands, especially with regard to church and state. Secondly, it focuses on how the clergy in both countries envisioned their role as knowledge broker and as pastor. Finally, it addresses the clergymen's activities in the fields of natural philosophy and agriculture and the nature of their devotion to knowledge accumulation and diffusion.

Church and State

Denmark and the Netherlands both have a long tradition of Protestantism. The Reformation turned Denmark in 1536 into a Lutheran state. Other religions were formally not admitted. The state church was an integral part

1 'de Videnskaber som er til virkelig Nytte for et Folk, de der skal hielpe Landet i trange Tider, de der skal frie Folket fra Armod, og Riget fra Undergang, de Videnskaber bør udbredes, de bør læres, drives paa og fortplantes, næst Guds Kundskab i alle Landets Kroge; og hvorved skeer det bedre, end naar Geistligheden, der boer i Hele Rigets Omkreds, faaer Materie til at arbeide baade Practice og Theoretice'. [Dyssel], *Tanker*, pp. 7–8.

of the state administration, and especially in the rural parts of the Danish conglomerate state, the clergy functioned as main representatives of the Crown. As a result, the Danish clergy was assigned an increasing number of formal tasks concerning registration and administration, health care, poor relief, and education in the parishes. This combination of the roles of pastor and civil servant required a regulated training at the University of Copenhagen. The broad role they had to fulfil in society was even reflected in the curriculum of the theologians. The future clergymen had to take for instance classes in agriculture, to prepare them for their role in the countryside. These men became the most important tool for the state in the religious and social disciplining of the people.[2]

In the Netherlands, the Reformed Church and its members had obtained a privileged position in society and in relation to the state in the sixteenth century. The close ties were, among other things, reflected in the opening up of political offices exclusively to members of this church as well as in the supervision of the church administration. Nevertheless, the Netherlands were a religiously pluriform country without a formal role for the clergy as civil servants. In 1809, for example, a large minority of the population (38.1%) was Catholic, the remaining majority comprised members of one of the protestant churches.[3] The Reformed Church was the largest denomination (55.5%), followed by the Lutheran (2.8%), Mennonite (1.4%), and Remonstrant churches (0.2%). Jews comprised 1.8 per cent of the population. During the short union of Belgium and the Netherlands between 1815 and 1830, the majority of the population was Catholic.

The religious pluriformity in the Netherlands did not discourage the state around 1800 from assigning the churches and its men a specific societal role.[4] In the spirit of Enlightenment thought, the proclamation of the Batavian Republic in 1795 and the forced flight of the ruling stadtholder William V, Prince of Orange, resulted in the declaration of freedom of religion, followed by decrees covering the dismantling of all the privileges of the Reformed Church in 1796 and the complete separation of church and state in 1798. Nevertheless, a concept for the 1796 constitution declared religion to be the foundation of the nation's happiness (*volksgeluk*). Already in the new constitution of 1801, the ties between church and state in the Netherlands were tightened. Especially the Kingdom of the Netherlands, established in 1814, was successful in exerting influence on the organisation of the churches

2 Jespersen, *A History of Denmark*, pp. 96–104.

3 Knippenberg, *Religieuze kaart*, p. 61.

4 Cf. Bijleveld, *Voor God, volk en vaderland*, esp. pp. 11–41.

and tried to strengthen a liberal theology. It strictly maintained an academic education as a condition to become a pastor, included classes in agriculture in the curriculum, and worked on installing a state salary for all the church men. For their daily living, Dutch pastors even then, like their Danish colleagues in this period, depended in part on the produce of their land. In 1815, King William I requested all pastors to incite the congregations to: 'virtue and godliness, attachment to King and Fatherland, and to that fraternal union which must guarantee future happiness'.[5] The newly established Dutch monarchy expected all churches to contribute to national unity.

About half of the two million people inhabiting the Netherlands belonged to the Reformed Church. While the Kingdom of Denmark was approximately the same area size as the Netherlands, it counted no more than one million inhabitants. Both countries had about the same number of parishes (the Netherlands approximately 1,500, Denmark circa 1,750); however, in Denmark they had a much stronger agricultural character. In the Netherlands cities were always close, even in the predominantly protestant rural areas of the north.[6]

Despite the differences in the historical relation between church and state as well as in the societal position of the churches, both countries assigned the clergymen an important role in the processes of state and nation building around 1800. In Denmark this role, however, was more extensive and formal, especially due to the long-existing and strong relation between the state church and the Crown. Both states shared the notion that the pastors were to contribute to the education of the people to a good and virtuous national community. As such, the changes in the formal Dutch policy in the nineteenth century regarding the churches have to be seen as a manifestation of what has been labelled as a broader, international perspective of Protestant Enlightenment.[7] Obviously, more was happening than only the state prescribing the clergy how to act.[8]

Educator of the People

Enlightenment thoughts were manifest in the Netherlands as well as in Denmark, predominantly in a moderate form stressing the moral

5 'deugd en Godsvrucht, gehechtheid aan Koning en Vaderland, en tot die broederlijke vereeniging, welke toekomstig geluk moet waarborgen'. Kist, *Neêrland's bededagen*, vol. 2, p. 630.

6 Bos, *In dienst van het Koninkrijk*, p. 101; Johansen, *Den administrative inddeling*.

7 Mijnhardt, 'The Dutch Enlightenment', pp. 198, 207. Cf. Edgren, 'The Enlightenment', pp. 63–64.

8 Jespersen, *A History of Denmark*, p. 102.

development of all inhabitants to virtuous citizens. Publicly expressed radical critique on the church was confined to individuals, even though we do not know what exactly was hidden under the surface since for instance the lifting of censorship in Denmark (1770–1773) resulted in fierce debates about the church.[9] However, in the eyes of many contemporaries, scientific progress did not necessarily conflict with theological ideas, as we can see from a poem from 1821 in the guestbook of the Franeker Planetarium:

> Planets, I see you here appointed circles
> (Imitated in miniature) completing your great journey
> How great you may be, how far from us located
> One Eisinga here follows your measured ways
> One Eisinga here shows the Creator's great splendour
> By what He has created from nothing in the universe
> You also show Eisinga by your ingenuity, & diligence
> By this masterpiece what a great man thou art.[10]

In general, the enlightened theological perspectives in both countries taught a less literal understanding of the scripture, a basic attitude that revelation did not conflict with ratio and to apply reason to nature or scripture as a means to obtain better understanding of God.[11] Changing theological opinions influence ideas about the role of the clergymen in all countries and periods.[12] In Denmark, the royal confessionary Christian Bastholm (1740–1819) published his thoughts about the clergy in 1794.[13] His main concern was how the profession was to become more useful and subsequently what the consequences were for the requirements of the pastors.

9 Cf. Langen and Stjernfelt, *The World's First Full Press Freedom*.

10 'Planeten, 'k zie U hier aangewezen Kringen / (in 't kleine nagebootst) Uw groten loop volbrengen / Hoe groot ge ook wezen moogt, Hoe ver van ons gelegen / Een Eisenga volgt hier Uw afgepaste wegen / Eén Eisenga toont hier des Scheppers groote pracht / Door 't geen Hij in 't Heelal uit niets heeft voortgebragt / Zo toont ge ook Eisenga door uw Vernuft & Vlijt / Door dit Uw Meesterstuk wat grote man gij zijt'. By M.D. Teenstra, 7 July 1821, in: *Gastenboek Planetarium*, p. 146. I thank Arjen Dijkstra for referring me to this quote.

11 For a full overview of Dutch and Danish church histories and theological perspectives in this period, see e.g.: Van Eijnatten and Van Lieburg, *Nederlandse religiegeschiedenis*; Schwarz Lausten, *Danmarks kirkehistorie*.

12 Cf. Clemens and Janse, *The Pastor Bonus*.

13 Bastholm, *Korte Tanker*.

His discussion about the best term for the clergy illustrates his ideas clearly. He argues in favour of the term 'people's teacher' (*folkelærer*):

> A people's teacher, a man whose vocation is to impart to his fellow-men religious as well as in human life useful common knowledge, whereby they are formed into virtuous Christians, useful citizens, and consequently into happy men.[14]

Bastholm stated that a pastor shows his congregation members the way to their eternal calling, but hopes that he can guide them in their temporal calling. This role depends less on command of the classical languages and more on the natural sciences. Bastholm suggests therefore to add chemistry, natural history, and economics to the curriculum that already consisted of classes in mathematics, physics, astronomy, and history. Becoming a useful people's teacher requires:

> Insight into physics, so that by explaining the manifold effects of nature by their true causes, he may, besides other important consequences, put an end to superstition: in chemistry, the knowledge of which has so great an influence in the proper treatment of the soil types, in dyeing, brewing, baking, churning, and the whole agriculture: in natural history, in so far as it has its influence on field, forest and garden-cultivation, giving knowledge of useful insects and their uses, and of the extermination of injurious insects, etc.: in veterinary science, to know how to treat domestic animals in their healthy and sick days: in mathematics, in so far as it is necessary for measuring, water drainage, useful tools' effective design for handling, etc.: in short, he must have proved by a public examination that he has a clear and thorough knowledge of all that pertains to a rural economy.[15]

14 'En folkelærer, en Mand, hvis Bestemmelse er at adbrede baade religiøse og i det menneskelige Liv almeennyttige Kundskaber blant sine Medmennesker, hvorved de dannes til dydige Christne, nyttige Borgere, og følgelig til lykkelige Mennesker'. Bastholm, *Korte Tanker*, pp. 11–12.

15 'Indsigt i Physiken, saa at han ved at udleede Naturens mangfoldig forskiellige Virkninger af deres sande Aarsager, hvorved, foruden andre vigtige Følger, Overtroen wil faae en Ende: i Chymien, hvis kundskab have saa stor en Inflydelse i Jordarternes rette Behandling, i Farveriet, Brygning, Bagning, Kiærning, og den hele Landhuusholdning: I Naturhistorien, for saavidt den haver sin Indflydelse paa Ager- Skove og Havedyrkning, giver Kundskab om nyttige Insekter og deres Anvendelse, og skadelige Insekters Udrøddelse o.s.v.: i Veterinær-Videnskaben, for at vide at behandle Huusdyrene i dere sunde og syge Dage: i Mathematiken, for saavidt den er nødvendig for Opmaalinger, Vandafledninger, nyttige Redskabers fordeelagtige Indretning i deres Haandteringer o.s.v.: kort, han maae ved een offentlig Prøve have beviist, at han haver en

In the Netherlands a number of pastors discussed the societal role of the clergy as well. The pastor occupied a prominent position within the nation thanks to his social importance, according to court preacher I.J. Dermout (1777–1867) in 1819.[16] The professor of theology at the University in Utrecht, Jodocus Heringa Ez. (1765–1840) stressed in the periodical *Church Counsellor and Adviser* (*Kerkelijke raadvrager en raadgever*) not only the ecclesiastical duties of the clergy, but also the importance of their so-called extraordinary activities (*buitengewone bezigheden*). They were described as:

> common activities, for the enlightenment, civilisation and care of the lower classes of the population, for the further training of more advanced people in useful science, for the establishment and maintenance of relations with people whose good character is legitimately valued.[17]

Heringa emphasised, like many contemporaries in Denmark and the Netherlands, the moral and civilising role of the teacher. One of his colleagues at the University of Groningen, the former pastor J.A. Uilkens (1772–1825), shared this view and explicitly addressed the need for the clergy to obtain and apply scientific knowledge. Much like Bastholm, Uilkens stated:

> regard the Teacher of Religion as the People's Teacher, who must disseminate all useful skills as widely as possible, and especially those which have the greatest influence on the prosperity of each *individual* and of society as a whole.[18]

The usefulness of science was a criterion for its importance and measured through its influence on individual and societal prosperity. Both countries struggled with a fear of economic decline, seeing it as the individual responsibility of each citizen to break the negative spiral. Some were better equipped for this patriotic duty than others. Thus, the bar was set high

tydelig og grundig Kundskab i alt det, som henhører til en Land-Oeconomie'. Bastholm, *Korte Tanker*, p. 29.

16 Dermout, *Leerrede*, pp. 13–14.

17 'gemeenschappelijke werkzaamheden, ter verlichting, beschaving en verzorging van de lagere volksstanden, ter verdere oefening van meer gevorderden in nuttige wetenschap, ter aanknooping en onderhouding van betrekkingen met menschen, op welker goede gezindheid men regtmatigen prijs stelt'. Heringa Ez., 'II', p. 69.

18 '[I]k beschouw den Leeraar van de Godsdienst als den Volksleeraar, die alle nuttige kundigheden zoo veel mogelijk moet verspreiden, en inzonderheid die, welke op de welvaart van iedere *individu* en van geheel de Maatschappij den grootsten invloed hebben'. Uilkens, 'Iets over de landhuishoudkundige lessen', p. 186.

for the clergy as educators of the people in the broadest sense, not only by the state but also by the foremen of the profession. The term was not new to designate the pastor, but here it is specifically linked to the idea of the enlightenment of the people.[19] It is therefore not surprising that a contemporary like the pastor Gerardus Benthem Reddingius (1744–1844) formulated his life motto as: '[I] have to work as long as it is day; night comes when no one can work'.[20] Despite this work ethic, he did not have enough time for all his endeavours. Instead of adjusting his workload, he requested assistance for his ecclesiastical work in order to keep up the extra activities.

Dissemination of New and Useful Knowledge

The extensive pleas to put theory into practice reflect what actually happened in many Dutch and Danish rectories. What discerns the pastor from most other educated people in the decades around 1800 is that the majority lived outside the large towns, just like the population. Here, the church and associated rectory formed the focal point of the pastor's daily work. From the pulpit he preached several times a week, in his rectory he studied and prepared his work, his shed functioned as a workshop, and on the accompanying land he not only grew crops to feed his family but had the opportunity to experiment and apply new scientific insights as well. We will explore these activities further.

Sermons and Smallpox

Preaching is regarded as one of the core tasks of the pastor. The number of sermons that have been published in this period are only a fraction of what has been preached in the churches. Nevertheless, the publications give us a clue as to what was delivered from the pulpits. Historians have shown that these sermons far from solely focus on evangelical content.[21] Daily life challenges, political questions, and disasters are recurrent themes. Of particular interest for this chapter is that some of the sermons focused on agricultural improvements, for instance the promotion of the use of potatoes – hence the term 'potato pastor' used by some historians. Allegedly, sometimes an individual pastor could be so devoted to sharing knowledge that he turned

19 Heitink, *Biografie van de dominee*; Roos, '"Jeg skriver for tænkende Mænd"', p. 132.

20 '[i]k moet werken zoo lang het dag is; de nacht komt, wanneer niemand werken kan'. Benthem Reddingius, *Herinnering*, p. v.

21 Cf. e.g. Bijleveld, *Voor God, volk en vaderland*; Bregnsbo, *Samfundsorden og statsmagt.*

a sermon about Matthew 4:4 ('Jesus answered [...] "Man shall not live on bread alone"') into an actual instruction on how to grow potatoes.[22]

Other illustrations of what and how knowledge could be disseminated from the pulpit are available. There are some sermons discussing smallpox, a deadly disease gradually brought under control in the course of the nineteenth century. In Denmark as well as the Netherlands people needed to be convinced to get inoculated and premiums were handed out to successful advocates of this method. The Dutch pastor Benthem Reddingius, from the small town Assen, published in 1817 his sermon titled *Lecture Delivered to Commend Cowpox Inoculation*.[23] It was held in reply to a call from the regional government and is quite classical in its approach. After having read the proclamation from the government, the bible text is read aloud, this text is explained, some comments are made, and finally the application is discussed. In the latter part, the bible text and the call for being treated are connected. With short references to Jenner as well as to the animal source of the serum, Benthem Reddingius presupposes broad knowledge about the technique among his audience and recommends it.

From Denmark we have another example, not only illustrating how likely it was that the clergy engaged in calls for going along with scientific progress, but also that the actual dissemination of new knowledge found a place in the pulpit. In 1803, pastor J.L. Hertel (1764–1831), who, according to the biographer of his son, was influenced by rationalism, produced an unusual but effective sermon.[24] The text entitled *On the Cowpox: A Sermon Preached in Rønnebæk and Ølstrup Churches on the Third Sunday after Trinity 1803* actually devoted a lot of space to explain the working of inoculation, then continued to refute all arguments against it from a religious perspective, and finally invited everyone to register with him if they wanted to receive a free treatment from the local doctor.[25] Comparable calls can be found throughout the eighteenth and nineteenth centuries and stimulated the vaccination rate, meanwhile illustrating the pastor's contribution to healthcare.[26]

Disease and Medicine

In approaches to eradicate smallpox we see some international differences in the role of the clergy. They possibly arise from differences in the relation

22 Bregnsbo, 'Kirkegangen og kirkekampen', p. 75. Reference to Brandt, *Potetprester.*

23 Benthem Reddingius, *Leerrede.*

24 Hertel, *Hans Wilhelm Hertel*, pp. 6–7.

25 Hertel, *Om Koekopperne.*

26 Roos, 'Medisinsk opplysning', pp. 23–37; Sandra, 'Sprokkelingen', pp. 2644–55.

of church and state as well as from population density. In Denmark the clergy was quite active in the campaign against smallpox and conducted about one quarter of the inoculations in the first years. One of the pastors who functioned as medic was Niels Blicher (1748–1839) who, when he had no medicine available, travelled 15 kilometres with two children to treat them with the pus from another child. By the time they developed blisters, he could use their pus to continue his medical work in the region. Blicher treated more than 1,000 people and taught the technique to others, including his son pastor Steen Steensen Blicher (1782–1848).[27] Whereas in Denmark these activities resulted in additional remuneration for the clergy, the focus in the Netherlands appears from the beginning to have been on medical professionals taking care of the vaccinations and pastors working to lower opposition to the method.[28] Given the role of the Danish clergy as civil servants, it seems logical to include them in these activities, especially since Denmark was more rural than the Netherlands and access to a doctor was more difficult.[29]

Despite the differences in the relation between state and church, we see that these Dutch and Danish pastors share a coherence between what was proclaimed and what was actually done. The conviction that new medicines and medication were important resulted in both countries in actions towards researching, developing, applying, and supporting methods to help cure diseases. New insights among the pastors could come from literature or acquired knowledge in academia, or arise in combination with an early form of scientific experimentation and research, very much in line with Enlightenment ideals.

In some cases, we see a prominent role for the clergy in the increase and spread of knowledge, for instance in veterinary medicine. The Dutch pastor Eelco Alta (1723–1798) dedicated himself in the province of Friesland to agriculture for – in his words – the good of the nation as well as for occasional diversion from studying theology.[30] Alta is especially remembered as the author of the piece that heralded the apocalypse with the conjunction of the planets and the sun in 1774 which, until recently claimed, reportedly triggered Eisinga to design his Planetarium.[31] It was based on empirical observations, just like the weather observations he

27 Bonderup, 'Reverend Blicher's Difficulties with Vaccinations', pp. 386–91.

28 Boschma, 'De bestrijding van de pokken', pp. 68–78.

29 Cf. Nørr, *Præst og administrator*, p. 391.

30 Cf. Breuker, 'Aacht maie 1774', pp. 33–36.

31 [Alta], *Philosophische bedenkingen*. Cf. the contribution by Arjen Dijkstra to this volume.

started making from 1744 onwards. From 1750 until at least 1792 daily observations were made with weatherglass and thermometer, which were used for publications in newspapers and possibly resulted in weather forecasts in almanacs as well. During most of his life as a pastor, Alta held cattle near the rectory. Here, he experimented with the inoculation against cattle diseases, published about his findings, treated hundreds of calves, and provided others with vaccines. This contribution was made possible thanks to his skills and access to cattle, as well as to the time gained from cancelling services due to the lack of churchgoers.[32] His role in improving veterinary medicine was profound and acknowledged already in 1780 when the Society for the Advancement of Agriculture (*Maatschappij tot bevordering van de Landbouw*) in Amsterdam awarded him a financial reward of 30 gold ducats.[33] His colleague pastor from the Remonstrant Church, Cornelis Nozeman (1720–1786), was actually the first to publish about these kinds of experiments in 1755. These and other pastors played a prominent role in the battle against veterinary diseases as well as in establishing national policy and educational institutions in the Netherlands.[34]

We find a Danish counterpart to Alta and his veterinary work in the person of pastor Peder Thun (1708–1783).[35] During his period in Hjortlund and Kalvslund, the rural areas in the western part of Jutland, he proved to be a frontrunner in the practice of inoculation. He claims to have been the first in Denmark to have inoculated children – his own six children – and was the first to inoculate against the cattle pest in 1762 with body fluids from sick cattle. His report about his work was published and lauded by the Royal Danish Agricultural Society (*Det Kongelige Danske Landhusholdningsselskab*). The Agricultural Society decided to establish an award for additional experiments and improved proof according to what was presented as 'Mr Pastor Thun's Method of Inoculation' (*Hr. Pastor Thuns Indpodnings-maade*).[36]

Science in the Countryside

Pastor Johan Arndt Dyssel rightly pointed out that the knowledge acquired by the clergy could be made to use thanks to the nature of their profession.

32 Talsma and Rommes, 'Dominees en de runderpest', p. 359.

33 Alta, *Verhandelinge*; Alta, *Nodige raadgeevingen*; Breuker, *Alta*, p. 245.

34 Talsma and Rommes, 'Dominees en de runderpest', pp. 356–63.

35 Paulli, 'Peder Thun'.

36 Thun, *Efterretninger*.

The pastors were shattered over the countryside and had a close connection to the congregation. He stated:

> What then was the use of young students [in theology] to have courses in physics and geometry, to hear lectures in botany, medicine, economics, and so on, if those whom Providence sends out as leaders of the common people were immediately condemned to forget both the sciences of which they had acquired a taste, and to be excluded from the useful nature, the workshop in which these are to be practised.[37]

Dyssel described himself as 'first celebrated as a poet, later as a practitioner of natural history and agricultural economics'.[38] The pastor, stationed on the island Lolland most of his life, was primarily occupied with agriculture. According to his biographer Bjørn Kornerup, he was a forerunner in promoting land replacement and in improving cultivation methods. He published on a number of topics and was awarded several medals by the Agricultural Society for his achievements.[39]

The situation in Denmark stimulated and facilitated the pastors to engage in knowledge advancement outside the ecclesiastical duties. The academic study of theology was preceded by two years of study at the faculty of philosophy. Here, at the end of the eighteenth century, future pastors were taught mathematics, astronomy, physics, history, and geography. The idea of the usefulness of natural sciences had become so prominent that the talented H.C. Ørsted (1777–1851), who was the first to prove the connection between magnetism and electricity, in 1801 received a three-year travel scholarship to obtain knowledge about how to make practical use of the natural sciences. Upon return, he received a position as lecturer at the university. Due to his popularity he was awarded a professorship in 1806 and taught physics and chemistry. For pastors the more practical knowledge was relevant since they depended for their daily living on the proceeds of their own land as well as on the revenues of the land of the parish members (Illustration 10).

37 'Hvad nyttede det da de unge Studerende, at høre Collegier i Physica og Geometrie, høre Forelæsninger i Botanik, Medicinen, Oeconomie og videre, naar de, som Forsynet sender ud som Ledere for Almuen, skulde strax dømmes til, at forglemme inden fire Begge de Videnskaber som de havde faaet Smag paa, og udelukkes fra den duelige Natur, det Værksted hvor disse skal udøves'. [Dyssel], *Tanker.*

38 'først bekiendt som Poet, bliver det siden som Practicus i Natur-Historien og Land-Oeconomien'. [Dyssel], *Tanker*, p. 11.

39 Kornerup, 'Johan Arndt Dyssel'.

Illustration 10. Stone set in Vilslev (Denmark) in 1953 by the parishes Vilslev, Darum, and Hunderup in memory of the work of pastor Knud Lang (1744–1781) for his contribution to agricultural reforms. Photo by Henning Pedersen

But more was at stake, as illustrated by Dyssel's appraisal of his colleague Philip Ernst Lüders (1702–1786) as a pastor who made use of science to strengthen the welfare of the realm and the wellbeing of its inhabitants.[40]

40 [Dyssel], *Tanker*, pp. 8–10.

Lüders started as court priest in Glücksborg in 1730, part of the Duchy Schleswig under the Danish monarch, later as provost in the same region. He is remembered as a prominent agricultural reformer. As new crops were introduced from the Netherlands and England, he tested them on his own land and tried to have them accepted among the farmers. He regularly ordered potatoes from the Netherlands, which he divided in small portions and spread among the peninsula Jutland in order to promote their use. He tested new farming methods, argued for the enclosure of fields and the cultivation of moors and heaths. For further dissemination of his knowledge, he made use of newly established journals in which he published extensively. He also tied in with the period's new way of spreading knowledge: he established in 1762 the Royal Danish Agricultural Society (*Königlichen Dänischen Ackerakademie*). This society received royal support and counted among its members many prominent names from the capital on the other side of the country. Relatively many pastors had signed up as well.

Other examples from Denmark are found in the work by Oluf Andresen Borreby (1737–1791), who, stationed as a pastor on the Danish island Bornholm, published *Instruction for the Practical Agricultural Cultivation, Based on Own Experience* in 1772.[41] He stressed that he wanted to contribute with the agricultural knowledge he had obtained on the farm of his parents, the knowledge he had gained at university and the experiences of putting theory into practice in his ecclesiastical community.[42] He was co-founder of *Íslenska akuryrkjufélagið*, a society for agriculture on Iceland, and shared his experiences as a farmer for others to benefit from. Almost half a century later, colleague pastor Niels Oschatz Rützow (1793–1846) submitted to the Agricultural Society a report of his own experiences with stable feeding in the summer and alfalfa growing at his rectory in Visby on the island Funen.[43] The Society had requested information about agricultural developments of importance. For this report, Rützow was awarded his second gold medal.

Besides sharing the results of agricultural experiments and experiences among peers and parish members, many pastors were involved in further distributing knowledge obtained through others. For instance, in the southern part of the island Sjælland, Lago Matthias Wedel (1752–1827) published between 1792 and 1796 a collection of five volumes on agriculture and land cultivation. It contained about 1,500 pages with summaries and excerpts from Danish and international journals aimed at pastors and others living

41 Borreby, *Anviisning*.

42 Borreby, *Anviisning*, p. 3.

43 Rützow, 'Hr. Pastor Rützows Beretning', pp. 125–46.

in the countryside.[44] In 1796 he even published an appendix of more than 350 pages entitled *Economic-Patriotic Treatises and Proposals Dedicated to Practical Farmers.*[45]

Wedel was well aware of what was taking place in the outlying areas. He travelled the Kingdom of Denmark and the Duchies Schleswig and Holstein and wrote his experiences and impressions down in *L.M. Wedel's Domestic Journey through the Most Significant and Beautiful Parts of the Danish Provinces*. This two-volume report offers a wealth of information about agriculture, economy, geography, folklore, and dialects. He visited some of the parishes covered in this chapter and relates the quality of the rectory's land in Nustrup to the outstanding work of, among others, Dyssel, and the knowledge of the farmers in Angeln, to the instructions and lessons of the 'the dignified, the memorable' Lüders.[46]

In the same decades in the Netherlands, we see similar initiatives that served the same goal: to contribute to the welfare of the state. The Dutch pastors could rely less for their role on their formal education though. Up until 1815, the church and not one of the universities (except for in Groningen) was in charge of examining the candidate-pastors. As a result, the focus was primarily on preaching, theology, and languages. Nevertheless, most pastors were educated at university. The situation changed as of 1815, when the state stipulated that an academic training was compulsory for new pastors and had to include more general knowledge like physics, astronomy, and agriculture. A broad *Bildungsideal*, as well as offering pastors the ability to improve the situation in the congregation, lay at the foundation of these plans.[47] Not less relevant, as in Denmark and other countries, was the fact that this knowledge should help improve the living standards in the countryside.

These ideas were broadly accepted in Europe, as we can read in one of the contributions to the Dutch periodical of the Society for the Advancement of Agriculture. This society was established in 1776 by prominent citizens to stimulate the applied scientific research on agriculture by organising annual competitions. The gold medal winner from 1786 praised Germany, England, and other countries for 'classifying agriculture among the sciences'.[48] The author of the piece, Joachim Fredrik Muller, citizen of Amsterdam, argued

44 Wedel, *Samlinger.*

45 Wedel, *Oeconomisk-patriotiske Afhandlinger.*

46 Wedel, *Indenlandske Rejse*. References to Dyssel in vol. 2, pp. 102–3, to Lüders in vol. 2, p. 46 (original qoute: 'den værdige, den uforglemmelige').

47 Bos, *In dienst van het Koninkrijk*, pp. 168–71; Van Eijnatten and Lieburg, *Nederlandse Religiegeschiedenis*, p. 232.

48 Muller, 'Verhandeling', pp. 7–13.

in favour of including agricultural knowledge in the academic curriculum and making it available for the clergy to contribute to the benefit of the commonwealth in general and the farmers in particular. His ideas were in that respect well ahead of what King William I tried to realise from 1814 onward. He added to his publication a detailed example of an 'agricultural catechism' that could be used in schools to increase pupils' knowledge in this field. The silver medal was awarded to Jan Willem van Loo (1759–1787), pastor in 's Heer Arendskerke. He may have been inspired by his uncle, pastor in a village congregation near Kleve, Germany, who educated him in ancient languages, German literature, and natural history before he went on to study theology at the university in Leiden, the Netherlands.[49] In his introductory remarks, he stressed that the Society could stimulate the village pastor (*Land- of Dorp-Leeraars*) to turn sermons into postils that present farmers the excellence, usefulness, and necessity of agriculture and animal husbandry in a clear, pleasant, and striking way.[50] Van Loo as well as Muller shared the enlightened ideal of the Society that education formed the basis for prosperity.

In the publications from the Society we find contributions from various countries authored by people from different layers in society, among others from the clergy. Bernhardt te Gempt (1792–1864), for instance, a pastor in Batenburg and renowned for his knowledge on natural philosophy, astronomy, literature history, and agriculture, was rewarded with a gold medal for his contribution on forestry.[51] Among the other authors, we find pastor Lambertus Aartsen (1750–1808), who was awarded medals for his studies on flax farming, calf diseases, and fertilising farmland. Pastor Leendert Calf from Haringhuizen, two of whose answers received reward, argued that the Society offered him the opportunity to share more than twenty years of agricultural experience to serve the patriotic farmer.[52]

The furthering of individual and societal prosperity could be based on the knowledge acquired at university or in other educational settings, but also on the insights gained through reading periodicals as well as participating in societies and conducting experiments. Science was put to use by, among others, the Dutch pastor J.H. Nieuwold (1739–1812), who according to contemporaries was renowned for the agricultural innovations that he tested and developed in his rectory garden close to the city Leeuwarden.

49 Van der Aa, 'Jan Willem van Loo'.
50 Van Loo, 'Verhandeling', pp. 126–28.
51 Fangman, 'Levensberigt', pp. 339–44.
52 'om den Vaderlandschen Landbouwer, van dienst te zyn'. Calf, 'Antwoord', p. 57.

He immersed himself in the latest literature and theories on agriculture in his study and tried to apply and improve them on his own ground. It was said that he was so successful that a small potato plant in his garden produced as many as 1,452 potatoes, a cauliflower plant produced so much seed that it could fill a wine bottle, and that experienced smokers could not distinguish the tobacco he grew from alien varieties. Critics claimed that Nieuwold wanted to get more out of nature than God had put into it. His more successful trials, including the new method of digging deeper and earthing crops, were emulated in the region. He also introduced the crop chicory, improved tools for harvesting it, and spread the custom of roasting the roots of the crop as a coffee substitute. In addition, he made proposals for moorland reclamation to increase the agricultural area. He was motivated by the idea that in this way he brought prosperity to his church members, thereby increasing the effect of his educational activities.[53]

One of his somewhat younger colleagues, J.A. Uilkens, worked ferociously from 1796 in the northern province of Groningen on educating and enlightening the countryside.[54] He founded a local department of one of the largest societies in the Netherlands (*Maatschappij tot Nut van 't Algemeen*) and presented, during its regular meetings in the region, reflections on nature, agricultural information from foreign journals, and physical questions. He experimented with new techniques (Illustration 11) and issued an almanac with a broad variety of topics covering advice on how to prevent spontaneous outbreaks of fire in corn, the possibility of using duckweed as pig food, and how to plant potatoes. He made knowledge accessible to a broad audience in publications, but also as an itinerant lecturer. He travelled the countryside to explain the functioning of the lightning rod with the help of a so-called thunder-church (*donderkerkje*). With the help of a strong discharge, a spark was generated that hit the church. With a lightning rod in place nothing happened, without the rod some gunpowder exploded and the church fell apart. Thanks to these experiments the number of installed lightning rods in the region was relatively high.[55]

Uilkens was motivated to contribute to the 'glorification of the great Creator through study of His creation, – the expansion of skills in this so useful and pleasant science, – the ennoblement of hearts through pure virtue

53 Visser, *Herinneringen*, esp. pp. 84–90. Based on Bijleveld, *Voor God, volk en vaderland*, pp. 172–73.

54 Based on: Bijleveld, *Voor God, volk en vaderland*, pp. 170–75.

55 Smit, 'Donder en bliksem', pp. 56–57. Cf. the contribution by Oliver Hochadel to this volume.

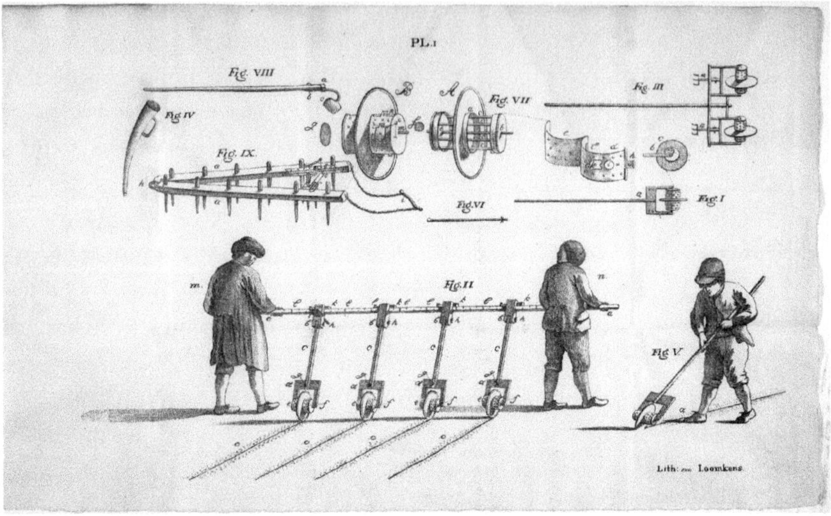

Illustration 11. Shows the 'sukereikroadsje' as improved by pastor J.A. Uilkens and used to sow rapeseed in rows. Lithograph by Jan Oomkens from A. Numan, *Iets voor landbouwers [...]* (Groningen 1827). Photo by Dirk Fennema (*Collection University Library Groningen, 'PS h 27*)

and godliness'.[56] His work and its impact was acknowledged when he was offered the new chair of agricultural economics in the faculty of mathematics and physics at the University of Groningen in 1815. The Mennonite pastor Jan Kops (1765–1849) was to become the professor in agricultural economics and botany in Utrecht. He is still renowned for having started the first scientific inventory in words and images of indigenous plants, entitled *Flora Batava*, which in approach and thoroughness bears great similarity to the standard work on the native bird species: *Nederlandsche vogelen* initiated by pastor Nozeman.[57] Only the chair of agricultural economics in Leiden was not occupied by a pastor. The three chairs had been established in 1815 to contribute to the agricultural knowledge of the clergy as means to improve the position of commoners in the countryside.[58] It constitutes an affirmation of what was already practice among many pastors. The appointment of Uilkens and Kops serves not only as recognition of the quality of their work, but also points out that the clergy produced knowledge as well.

56 'verheerlijking van den groten Schepper door de beschouwing van zyne schepping, – de uitbreiding van kundigheden in deze zo nuttige en aangename wetenschap, – de veredeling des harten door reine deugd en godsvrucht'. Uilkens, *De volmaaktheden*, vol. 1, 'voorberigt'.
57 *Flora Batava* was issued between 1800 and 1934, and edited and republished in 2023: Van Gelder and Peeters, *Flora Batava*. Nozeman, et.al., *Nederlandsche vogelen* appeared between 1770 and 1829, and was reissued in 2014.
58 Van Berkel, *Universiteit van het Noorden*, vol. 1, p. 526.

Knowledge Brokers

The prominent position of Danish and Dutch pastors in societies as founders, members, authors, and prize winners illustrates the central role they occupied as knowledge brokers in national and even international networks. Knowledge obtained during university studies, reading literature, and conducting research was shared through ecclesiastical activities, best practices, public and academic lectures, publications, and societies. Whereas this chapter has focused on topics relating to natural philosophy and agriculture, we must keep in mind that pastors could be devoted to many other subjects as well.[59] In every ecclesiastical tradition church and school are closely connected, for instance. Not only because of the religious education of the youth, but also because the clergy often contributed to the education of the schoolmaster as well as the many publications on pedagogical topics by Danish and Dutch pastors. As the primary school gained a more prominent role in Enlightenment thought as an important instrument in the raising of good citizens, pastors remained involved into the nineteenth century. In Denmark this responsibility was one of the pastor's tasks as a civil servant, whereas the Dutch government appointed citizens to become school inspectors, which in the first decades of the nineteenth century often were pastors.[60] The pastor Nieuwold was lauded for his contribution as the Dutch Pestalozzi, a reference to the famous Swiss pedagogue. For all involved, school was an important instrument in enlightening the people. In both countries around 1800, a sense of decline prevailed. In Denmark there was particular concern for the moral state of the farmers, whereas in the Netherlands many felt that the decline of the seventeenth-century Republic was connected to a moral decline. In both countries efforts aimed at enlightening the people by advancing, sharing, and applying useful knowledge in and outside church.

Devoted to Useful Knowledge

This chapter has highlighted several Danish and Dutch pastors who were dedicated to and successfully propagated and disseminated knowledge that was not part of the core of their profession. They aimed for knowledge advancement and dissemination, ranging from developing and applying

59 Cf. Bijleveld, *Voor God, volk en vaderland*, pp. 109–83; Bijleveld, 'Language, National Culture and the Clergy'.

60 Cf. Bijleveld, *Voor God, volk en vaderland*, pp. 146–59; Nørr, *Præst og administrator*, pp. 27–228.

the latest insights in science to conducting research and disseminating its results. These activities transcended the boundaries of the profession and qualifies many pastors as devotees of science. By practising science, they wanted to contribute to individual and societal prosperity. From a theological perspective, putting science to use strengthened the ecclesiastical message and contributed to the knowledge and glorification of the creator and his creation. Thus, there was a fruitful interaction between clergy and science among the protagonists of a more moderate Enlightenment.

There are numerous publications by clergymen that were issued by societies, appeared in journals of societies, or were a response to organised competitions by societies. To some extent, the overarching objectives of these societies may have incentivised the clergy to delve into certain topics. The fact that pastors depended partly on the proceeds of the land for their livelihood may also have played a role. But a closer look at the organisations reveals that the general goal of these societies to share (practical) knowledge and thus contribute to the welfare of the state coincides with the clerical ideals in this period. It is therefore not surprising that we find pastors among the founders of societies, as well as in some cases an overrepresentation of the clergy among the members and submitters of responses to contests. These organisations and their ideals coincided with the calling many pastors felt. The societies offered a platform for what Peter Burke described as the new ideal of the eighteenth- and early nineteenth centuries, the 'Man of Letters'; an individual who wrote poems, plays or novels, made contributions to the humanities and showed an interest in natural sciences.[61] The societies offered a network of likeminded people connecting science and practice, cities and countryside. Many pastors approached this ideal thanks to their vocation since it offered them the required knowledge, means, and network. They were part of an international phenomenon.

In the course of the nineteenth century the humanities and sciences drifted apart and the professionalisation and specialisation of science in universities complicated the possibilities of excelling in one or more disciplines outside established institutes.[62] Pastors were less likely to obtain a professorship outside theology by that time. Parallel to this, in the Netherlands and Denmark the call that had always been present in the church to focus more on ecclesiastical topics than on practical and natural sciences gained strength. For instance, in both countries the proportion of

61 Burke, *The Polymath*, pp. 87–88.

62 Burke, *The Polymath*, p. 119. Snelders, 'Professors, Amateurs, and Learned Societies', p. 323. Cf. the contribution in this volume by Klaas van Berkel.

agricultural knowledge in the curriculum at universities was reduced. The days of the pastor as a devotee of science were numbered.

This comparison between pastors of the Danish Lutheran Church and of the Dutch Reformed Church illustrates that the clergy's activities were typical neither of state nor of denomination; nor did they necessarily depend on the constitutional position of the church. Unfortunately, there are not enough sources to study a representative sample of pastors to be able to draw conclusions on the profession as a whole, especially not since this chapter has focused on a specific group of pastors and their involvement in knowledge advancement. It would be highly relevant to further explore how these activities outside church permeated, for instance, regular church work like preaching to get an indication of the actual impact of the pastor's devotion to science. Further research could investigate the differences and similarities between Catholic and Protestant clergy, thus contributing to uncovering the importance of the social position of the individual pastors, of an academic background and bourgeois network.

Some variations between the Lutheran and Reformed pastors existed and appear to derive from, besides personal interest, the relation between state and church, as well as national economic activities and the population density. The similarities, regardless of protestant denomination, had ramifications on domestic as well as international developments. Both countries shared ideals as well as a great concern for future prosperity, for which the enlightenment of the people served as the solution for the country. Individual elevation was therefore a patriotic duty. The clergy's actual activities beyond the strictly ecclesiastical work contributed to knowledge advancement and dissemination in combination with moral improvement, which tied in with contemporary Enlightenment ideals and structures. Thanks to their position, education, and network the pastors could participate in this international phenomenon of devotees of science.

Bibliography

Alta, E., *Verhandelinge over de natuurlyke oorzaaken der ziekte onder het rund-vee en derzelver langere duuringe als te vooren, waar in* [...] *de in-enting derzelve* [...] *wordt aangepreezen: Gaande vooraf twee vertoogen, gezonden aan de Hollandse Maatschappy der Weetenschappen, ter beantwoording eener vraag over deeze stoffe in 1759 voorgesteld en in 1760 herhaald* (Leeuwarden 1765).

Alta, E., *Nodige raadgeevingen aan overheden en ingezetenen, dewelke in het bijzonder voor den boer in deeze akelige omstandigheden, waarin het vaderland zig door de ziekte van het rundvee thans bevind, van zeer veel nut kunnen zijn* (Leeuwarden 1769).

[Alta, E.], Liefhebber der Waarheid, *Philosophische bedenkingen over de conjunctie van de planeten, Jupiter, Mars, Venus, Mercurius en de Maan, op den 8en May 1774 staande te gebeuren, en wel over de mogelijke en waarschijnlijke sterre- en natuurkundige gevolgen deezer conjunctie* (Leeuwarden 1774).

Bastholm, C., *Korte Tanker til nærmere Eftertanke over Den geistlige Stand* (Copenhagen 1794).

Benthem Reddingius, G., *Leerrede, gehouden tot aanprijzing der koepok-inenting* (Assen 1817).

Benthem Reddingius, G., *Herinnering der opleiding tot de evangeliedienst en der ondervinding in de (bijna) vijftigjarige waarneming daarvan, nagelaten door G. Benthem Reddingius* (Assen 1844).

Bijleveld, N.H., *Voor God, volk en vaderland: De plaats van de hervormde predikant binnen de nationale eenwordingsprocessen in Nederland in de eerste helft van de negentiende eeuw* (Delft 2007).

Bijleveld, N.H., 'Language, National Culture and the Clergy in Nineteenth-Century Denmark', in: P. Broomans et al., ed., *The Beloved Mothertongue: Ethnolinguistic Nationalism in Small Nations: Inventories and Reflections* (Louvain 2008), pp. 93–107.

Bijleveld, N.H., 'The Nationalization of Christianity: Theology and Nationalism in Nineteenth-Century Denmark', *Tijdschrift voor Skandinavistiek* 31, no. 2 (2010), pp. 77–97.

Bonderup, G., 'Reverend Blicher's Difficulties with Vaccinations in 1807', *Dansk Medicinhistorisk Årbog* 39 (2011), pp. 35–40.

Borreby, O.A., *Anviisning til den practiske Agerdyrkning, grundet paa egen Erfaring* (Copenhagen 1772).

Bos, D.J., *In dienst van het Koninkrijk: Beroepsontwikkeling van hervormde predikanten in negentiende-eeuws Nederland* (Amsterdam 1999).

Boschma, C., '"Een' pest, die Vrieslands Steeden, en luchtgewest bedorf": De bestrijding van de pokken', *De Vrije Fries* 64 (1984), pp. 68–78.

Brandt, N., *Potetprester* (Oslo 1973).

Bregnsbo, M., *Samfundsorden og statsmagt set fra prædikestolen: Udviklingen i præsternes syn på samfundsorden og statsmagt i Danmark 1750–1848, belyst ved trykte prædikener* (Copenhagen 1997).

Bregnsbo, M., 'Kirkegangen og kirkekampen: Udfordringerne til kirke og gejstlighed i de sidste årtier af 1700-tallet og den danske præstestands håndtering af dem', *Sjuttonhundratal* 3–4 (2006/2007), pp. 66–84.

Breuker, Ph.H., 'Aacht maie 1774: Panyk en Ferljochting', *De Vrije Fries* 64 (1984), pp. 26–46.

Breuker, Ph.H., *Eelko Alta (1723–1798): Biografie van een verlicht volksvoorlichter* (Amersfoort 2024).

Burke, P., *The Polymath: A Cultural History from Leonardo da Vinci to Susan Sontag* (New Haven 2020).

Calf, L., 'Antwoord op de vraag [...]: Welke wyze is de beste, om de verschillende soorten van bouwgronden in ons vaderland [...] tot eene grooter en duurzaamer vrugtbaarheid te brengen?', *Verhandelingen uitgegeeven door de Maatschappy ter bevordering van den landbouw te Amsterdam* 3, no. 2 (1784), pp. 54–78.

Clemens, T., and W. Janse, ed., *The Pastor Bonus: Papers Read at the British-Dutch Colloquium at Utrecht, 18–21 September 2002* (Leiden 2004).

Dermout, I.J., *Leerrede ter aanprijzing van het leeraarsambt, door I.J. Dermout, Buitengewonen hopprediker van z.m. den koning der Nederlanden, predikant in 's Gravenhage en secretaris der sijnode van de hervormde kerk: Uitgegeven op last van de sijnode* (The Hague 1819).

[Dyssel, J.A.], Philander, *Tanker om Præsternes Gaarder og Jorder samt disses Indvikling i det skadelige Fælledskab* (Copenhagen 1771). https://tekster.kb.dk/text/tfs-texts-1_001-shoot-workid1_001_012 [accessed 29 April 2024].

Edgren, H., 'The Enlightenment', in: M. Roos et al., ed., *Exploring Textbooks and Cultural Change in Nordic Education 1536–2020* (Leiden 2021), pp. 63–65.

Engelhardt, J., *Borgerskab of fællesskab: De patriotiske selskaber i den danske helstat 1769–1814* (Copenhagen 2010).

Fangman, H., 'Levensberigt van Bernhardt te Gempt', *Jaarboek van de Nederlandse Maatschappij der Letterkunde* (1864), pp. 339–44. https://www.dbnl.org/tekst/_jaa002186401_01/_jaa002186401_01_0032.php [accessed 29 April 2024].

Gastenboek Planetarium, vol. 2 (Royal Eise Eisinga Planetarium).

Heitink, G., *Biografie van de dominee* (Baarn 2001).

Heringa Ez., J., 'II. De onderscheiden bezigheden aan het Prediksambt verbonden', in: *Kerkelijke raadvrager en raadgever* 1, no. 1 (1821), pp. 61–134.

Hertel, J.L., *Om Koekopperne: En Prædiken, holden i Rønnebæk og Ølstrup Kirker paa tredie Søndag efter Trinitatis 1803* (s.l. [1803]).

Hertel, L., *Hans Wilhelm Hertel: Bidrag til den sønderjydske Folkevækkelsens Historie* (Odense 1897). https://slaegtsbibliotek.dk/928217.pdf [accessed 29 April 2024].

Jespersen, K.J.V., *A History of Denmark* (Basingstoke 2011).

Johansen, H.C., *Den administrative inddeling*. https://danmarkshistorien.lex.dk/Den_administrative_inddeling [accessed 29 April 2024].

Kist, N.C., *Neêrland's Bededagen en Biddagsbrieven*, vols. 1–2 (Leiden 1848–1849).

Knippenberg, H., *De religieuze kaart van Nederland: Omvang en geografische spreiding van de godsdienstige gezindten vanaf de Reformatie tot heden* (Assen 1992).

Kornerup, B., 'Johan Arndt Dyssel', *Dansk Biografisk Leksikon*. https://biografiskleksikon.lex.dk/Johan_Arndt_Dyssel_-_præst [accessed 29 April 2024].

Langen, U., and F. Stjernfelt, *The World's First Full Press Freedom: The Radical Experiment of Denmark-Norway 1770–1773* (Munich 2022).

Mijnhardt, W.W., 'The Dutch Enlightenment: Humanism, Nationalism, and Decline', in: M.C. Jacob and W.W. Mijnhardt, ed., *The Dutch Republic in the Eighteenth-Century: Decline, Enlightenment, and Revolution* (Ithaca 2019), pp. 197–223.

Muller, J.F., 'Verhandeling over de samenstelling van een landbouwk: Schoolboek', in: *Verhandelingen uitgegeeven door de Maatschappij ter bevordering van den landbouw te Amsterdam* 5, no. 1 (1786), pp. 3–119.

Nozeman, C. et al., *Nederlandsche vogelen*, 1770–1829 (reprint: Tielt 2014).

Nørr, E., *Præst og Administrator: Sognepræstens funktioner i lokalforvaltningen på landet fra 1800 til 1841* (Copenhagen 1981).

Paulli, R., 'Peder Thun', *Dansk Biografisk Leksikon*. https://biografiskleksikon.lex.dk/Peder_Thun [accessed 29 April 2024).

Roos, M., '"Jeg skriver for tænkende Mænd, der selv maae kunne prøve Aanderne": Nikolai Lorenz Fallesens tidsskrift for religionslærere (1793–1808)', *Sjuttonhundratal* 10 (2013), pp. 126–44.

Roos, M., 'Medisinsk opplysning og sosial mobilisering fra prekestolen – Presten som offentlig helse- og sosialarbeider i Danmark-Norge mot slutten av 1700-tallet', *Teologisk tidsskrift* 4 (2015), pp. 23–37.

Rützow, N.O., 'Hr. Pastor Rützows Beretning til Landhuusholdningsselskabet om hans Sommerstaldfodring og om hans Avl af Lucerne; m.v.', *Nye Landoekonomiske Tidender* 10 (1828), pp. 125–46. https://tidsskrift.dk/tidsskriftlandoekonomi/article/view/98161 [accessed 29 April 2024]

Sandra, H., 'Sprokkelingen betreffende de inenting tegen pokken in ons land omstreeks 1800', *Nederlands Tijdschrift voor Geneeskunde* 81 (1937), pp. 2644–55.

Schwarz Lausten, M., *Danmarks kirkehistorie* (Copenhagen 2004).

Smit, F., 'Donder en bliksem: Donderkerkje', in: F. Smit, *Panorama van de wetenschap* (Groningen 2022), pp. 56–57.

Snelders, H.A.M., 'Professors, Amateurs, and Learned Societies: The Organization of the Natural Sciences', in: M.C. Jacob and W.W. Mijnhardt, ed., *The Dutch Republic in the Eighteenth-Century: Decline, Enlightenment, and Revolution* (Ithaca 2019), pp. 308–23.

Talsma, E.J., and R. Rommes, 'Dominees en de runderpest in de tweede helft van de 18e eeuw', *Argos* 59 (2018), pp. 356–63.

Thun, P., *Efterretninger om Qvægsygens Indpodning, indsendte til Det kongelige danske Land-Huusholdings-Selskab, med Selskabets hosføjede Erindringer, og udsatte Extra-Premier, for nye Erfaringer og nøjagtigere Beviis paa Qvægindpodningens Nytte, efter Hr. Pastor Thuns Indpodnings-Maade* (Copenhagen 1770).

Uilkens, J.A., *De volmaaktheden van den Schepper in Zyne schepselen beschouwd, tot verheerlijking van God, en tot bevordering van nuttige natuurkennis*, vols. 1–4 (Groningen 1801–1823).

Uilkens, J.A., 'Iets over de landhuishoudkundige lessen aan onze Hooge-scholen en over het ook daarvoor uitgegeven landhuishoudkundig handboek', *De Recensent, ook der Recensenten* (*Antikritiek en Mengelwerk*) 14, no. 2 (1821), pp. 181–99.

Van Berkel, K., *Universiteit van het Noorden: Vier eeuwen academisch leven in het Groningen*, vols. 1–3 (Hilversum 2014–2022).

Van der Aa, A.J., 'Jan Willem van Loo', *Biographisch woordenboek der Nederlanden*, vol. 11 (1865). https://www.dbnl.org/tekst/aa__001biog13_01/aa__001biog13_01_1278.php [accessed 29 April 2024].

Van Eijnatten, J., and F. van Lieburg, *Nederlandse religiegeschiedenis* (Hilversum 2005).

Van Gelder, E., and N. Peeters, ed., *Flora Batava* (Tielt 2023).

Van Loo, J.W., 'Verhandeling over de samenstelling van een landbouwk: Schoolboek', in: *Verhandelingen uitgegeeven door de Maatschappij ter bevordering van den landbouw te Amsterdam* 5, no. 1 (1786), pp. 123–45.

Visser, H.W.C.A., *Herinneringen aan J.H. Nieuwold, door H.W.C.A. Visser* (Sneek 1814).

Wedel, L.M., *Samlinger om Agerdyrkning og Landvæsen: I Korthed uddraget af de beste oeconomiske danske og fremmede Skrifter til Brug for Landsbye-Præster og andre paa Landet boende*, vols. 1–5 (Copenhagen 1792–1796).

Wedel, L.M., *Oeconomisk-patriotiske Afhandlinger og Forslag praktiske Landmænd tilegnede, som et Tillæg til Samlinger om Agerdyrkning og Landvæsen* (Copenhagen 1796).

Wedel, L.M., *L.M. Wedels Indenlandske Rejse igjennem de betydeligste og skjønneste Egne af de danske Provindser*, vols. 1–2 (Copenhagen 1803).

About the Author

Nikolaj Bijleveld is managing director of the University of Groningen Business School.

Mediation of Knowledge

Artefacts and Context

8. The Oratorian Teodoro de Almeida (1722–1804) as a Planetarium Maker in France and Portugal

Luís Tirapicos and Samuel Gessner

Abstract: The Oratorian priest Teodoro de Almeida (Lisbon, 1722–1804) is best known for his multi-volume *Recreação filosófica* (1751–1799), a series of popular books on natural philosophy. His life as a devotee of science comprised, however, several other facets. The variety of Teodoro's activities is best understood in the context of his successive social conditions as he moved from the college of a religious congregation to urban society, to court and eventually to the realm of an academy of sciences. These different environments and audiences had repercussions on the consecutive conceptions and realisations of planetariums he conceived over the years. Teodoro's activity shows him continuously searching for ways to convey scientific knowledge to various audiences, especially those outside the traditional schools.

Keywords: Recreação filosófica, natural philosophy, astronomy, religious congregation, urban society, court

As a young man, the Portuguese devotee of science, Teodoro de Almeida, already admitted cleric of the Congregation of the Oratory, worked as an assistant to his teacher of experimental philosophy, in Lisbon. Soon after, he would himself occupy this chair at the Oratory's school in 1751. Political persecutions, however, lead to Teodoro's departure to Porto and later to his exile in Bayonne in France. There, Teodoro made a living from private lessons of Newtonian physics. As part of this activity, he conceived his first planetarium, an 'orrery type' mechanism, to demonstrate the heliocentric motion of the planets. Teodoro used to emphasise, then already, that his planetarium was 'better' than the one famously designed by Desaguliers (1683–1744).

Bijleveld, Nikolaj, Arjen Dijkstra, and Samuel Gessner, eds. *Cultivating and Communicating Natural and Technical Knowledge around 1800: Devotees of Science.*
Taylor & Francis Group, 2025
DOI 10.5117/9789463728577_CH08

Illustration 12. Portrait of Teodoro de Almeida. Oil on canvas by J.B. Gérard, 1801 (*Provided by Arquivo Nacional da Torre do Tombo, Lisbon*)

Eventually, financial reasons and his imminent departure from Bayonne pushed Teodoro to sell his teaching equipment at auction. Called back to Lisbon, as the Oratorian order was restored in Portugal, Teodoro became confessor to the Queen Dona Maria (1734–1816). He made a new planetarium, introducing additional improvements, this time recurring to the workshops of the army arsenal. He also included a tellurium (a moveable model of the Earth-Moon-Sun system to demonstrate the phases and eclipses) as a supplementary apparatus that could be stored in one of the drawers of the planetarium, a device that he presented to the Academy of Sciences of which he was a founding member. He dedicated the new planetarium to the

Queen's son, prince Dom João (1767–1826). In return, the prince offered more funds to produce even a further planetarium so that Teodoro would not be deprived of his teaching equipment. This one would include the 'novelty' of showing the ecliptic plane in the vertical position, like in a longcase clock.

This quick overview of the momentous trajectory of Teodoro de Almeida (Illustration 12) not only shows the clergyman's resilience but also highlights some remarkable traits of eighteenth-century astronomy. Firstly, religious congregations offered a space for the discussion of the various alternative world systems (Ptolemaic, Tychonian, Copernican) which contributed to the spread of the new views of astronomy. They also practised experimental philosophy. Secondly, members of the clergy, such as Teodoro, thrived in this context by combining their pastoral activity with the study and teaching of natural theology. Both faith and knowledge of nature are ways to praise God as the wise architect of the universe. Given the time investment, gained expertise, and impact of this work that seamlessly melted with his nominal occupation as a priest, it is correct to label Teodoro de Almeida a 'devotee of science'. Thirdly, scientific erudition, the defence of a 'modern' approach to nature (founded on reason and experiment), and communication skills using instruments including orreries had a political role to play. Teodoro like so many other clergy embraced it in a time when society transformed itself by enlightenment currents and the clergy, accordingly, was pushed to redefine its role, legitimacy, and 'usefulness'. These three points will receive exemplary illustration in the biography of Teodoro that follows.

Teodoro's case also prompts further questions. How did Teodoro find his role of devotee of science in the society of eighteenth-century Portugal and France? And we may ask, more specifically, what effect did the successive environments and audiences, which Teodoro encountered along his life, have on the form and content of his planetariums? This chapter examines whether the changing context has had repercussions on the successive conceptions and realisations of the instruments, as the clergyman moved from the school of a religious congregation, to urban society, to court, and eventually the Academy of Sciences.

The Successive Environments of Teodoro's Activity

Lisbon: Teaching at the Oratorian Congregation (1748–1760)

Teodoro's initial work in physics and astronomy happened in an utterly academic environment, although not at university level. He entered the Congregation of the Oratory of St Philip Neri, in Lisbon, as a 13-year-old boy. The Oratorians, an order founded by Neri (1515–1595) in Rome and confirmed by papal bull in 1575, organised communities without strict rules in multiple countries. The order focused on two areas of action: spiritual and material assistance to the poor and education and teaching. After three years of courses in philosophy with Father João Baptista (1705–1761) as physics teacher, he pursued another four years of study of theology. Teodoro – who had also studied mathematics in parallel – then became the teaching assistant of Father Luís José (1725–1808) in 1748. He would succeed him on the chair of physics in 1751.

Teodoro's early years fell into an agitated period as far as a Portuguese debate on knowledge and education was concerned. Luís José's predecessor on the physics chair, Father João Baptista, played an important role emphasising the difference in teaching between the Oratorians and the Jesuits in Lisbon. In his *Philosophia aristotelica restituta et illustrata qua experimentis, qua ratiociniis nuper inventis* (Lisbon, 1748) Baptista argued that in his congregation the approach of experimental philosophy was in agreement with the original work of Aristotle, and the apparent contradictions were the result of a bad reading of the philosopher's works by those whom he termed 'Peripatetics'. By this term Baptista implicitly targeted and disparaged the Jesuits.[1] João Baptista became not only an important advocate of the new experimental philosophy among the Oratorians but was also one of the first Portuguese scholars to publish and comment on Newton's physics, using staged experiments to illustrate the correctness of the English natural philosopher's findings.

To establish the Oratorian congregation at the palace and convent of Nossa Senhora das Necessidades in Lisbon, the king had provided funds, including for the equipment of a mathematical-physical cabinet. This was in line with the creation of new physics cabinets all over Europe in the eighteenth century, including those appearing in colleges such as the ones in Warsaw (Collegium Nobilium, 1740), Ingolstadt (Jesuit University, ca. 1748), Prague (Jesuit University, ca. 1750), and Dillingen (Jesuit University, ca. 1752) – to name just a few examples.[2] At this Oratorian Hospice, beyond the lectures for students (in Latin), Baptista started, in 1752, to organise conferences (in Portuguese), mostly attended by an interested public of court nobility and Lisbon notables, occasionally even by King Dom José himself.[3] It is well known that lessons on experimental physics, usually the reserve of

1 Pereira Gomes, 'João Baptista', pp. 121–37.

2 Wyka, 'Collections', pp. 173–93; Hellyer, *Catholic Physics*, pp. 181, 186–87, 198–201.

3 Contente Domingues, 'Um Projecto Enciclopédico', p. 242.

enrolled students, were opened up to a distinguished audience in various places in Europe.[4] Famously, Nollet's lessons in France attracted among others the Duke of Chartre, and even royalty including the king of Sardinia, the dauphin of France, and Queen Maria Leszczyńska.[5]

As a natural consequence of the close relation between Teodoro and his master João Baptista, Teodoro was chosen to present the experimental lessons to the select public during the summer months of 1753. As he had been working on a treatise on modern natural philosophy in Portuguese in dialogue form, showing Teodoro's interest and ability to present scientific content in pleasant prose, one might speculate that this made him an ideal candidate. Thus, these public demonstrations can be seen as a continuation of the promotion of experimental philosophy using the well-furnished cabinet of the Oratorian House of Necessidades.[6] Sometime later, Teodoro de Almeida would even be charged with teaching natural philosophy in the regular classes in the Oratorian House.

The same vision and convictions that drove the public lessons of experimental philosophy had possibly also motivated Teodoro's project on natural philosophy. In 1751 the first two volumes (of the planned six) of Teodoro's books on experimental philosophy were published. They revolved around mechanics and hydrostatics, whereas the later volumes ordered the topics of natural philosophy according to the themes of colour, sound, taste, heat, and cold. The full title of the volumes reads: *Philosophical Recreation or Dialogue among Theodosio, Eugenio and Silvio about Natural Philosophy for the Education of the Curious Who Did Not Attend the Schools*.[7] As the title of this work explicitly states, the imaginary prospective reader of the work would be someone 'who has not attended classes'. Thence, the choice of publishing in Portuguese was deliberate, and Teodoro does not seem to have published any work in Latin. His works echo the efforts to disseminate modern physics to a general, cultivated readership in vernacular languages as they were occurring throughout Europe with Desaguliers's *Mathematical Elements of Natural Philosophy* of 1731, Musschenbroek's *Beginselen der natuurkunde* of 1736, Algarotti's *Newtonianismo per le dame, ovvero dialoghi*

4 Schaffer, 'Natural Philosophy', pp. 1–43. Also in Italy (e.g. Rome, Genova) experimental physics lessons were held at specific times in Italian using the university collection of apparatus. See Leone et al., 'La Fisica', pp. 167–84; and Findlen, 'Science as a Career', pp. 441–69.

5 Taton, *Enseignement et diffusion*, p. 624.

6 Contente Domingues, *Ilustração*, pp. 40–43.

7 *Recreação filosófica ou dialogo sobre a filosofia natural para instrucção de pessoas curiosas, que não frequentarão as aulas, por Theodosio, Eugenio, Silvio*. Two volumes published in 1751, printed by Miguel Rodrigues.

sopra la luce e i colori of 1737, Nollet's *Leçons de physique expérimentale* (1743–1748), to name just a few.[8] In the Lusophone world Teodoro's work had no equal and it soon went through multiple editions. It was also translated to Spanish and distributed widely in Latin America.[9] The entire content of *Recreação filosófica* was presented in the form of a dialogue between three figures: Teodósio (the author's alter ego), Sílvio (a peripatetic, representative of the Aristotelian view of physics), and Eugénio (an engineer). Following an 'eclectic' approach the work clearly advocated the 'modern' (i.e. experimental) philosophy against the traditionalists' perspective.[10]

To understand Teodoro's motives, it is important to take a look at his particular social situation. In teaching and writing about natural philosophy Teodoro found his individual calling as an Oratorian. Over his long career he would stay faithful to this commitment. Here, a distinctive aspect of Oratorian organisation must be singled out. While the Oratorian congregation shared with other religious orders, such as the Jesuits, the objective of evangelising society, it did not impose fixed rules on its members. The ideal Oratorian was to be reactive to whatever needs or opportunities would appear. As a consequence, the congregation was open to the fresh directions in science, and was therefore uniquely suited to a life project such as the one pursued by Teodoro whose talent as a populariser of natural knowledge became ever more evident.[11] Becoming a devotee of modern science, in particular as its populariser, constituted for Teodoro the way of fulfilling his mission as an Oratorian.

Since the early stage of his career as the assistant of professor Luís José, Teodoro was familiar with the use of an orrery to demonstrate the heliocentric system. This may well have been the inspiration for his later undertaking of constructing such a device of his own. We learn from a late document *The Life of Padre Teodoro de Almeida* that an orrery was available in the House of the Oratorian Congregation, housed in the convent of Nossa Senhora das Necessidades.[12] It was an orrery conceived around 1730 according

8 See also Fiolhais, 'Os diálogos', p. 95.

9 Contente Domingues, *Ilustração* (pp. 69–70) mentions further works containing scientifically eclectic views, based on reason and experiment that circulated in Portugal. The *Compendio Mathematico* (1707–1715) by the Oratorian Tomás Vicente Tosca and the *Teatro critico universal* (1726–1739) by the Benedictine cleric Benito Jerónimo Feijoo, both written in Castillian language. Teodoro certainly knew both works.

10 Silva, 'Recreação Filosófica', pp. 206–18.

11 Dos Santos, 'Oratorianos', pp. 231–32, 237.

12 *Life*, § 75. This anonymous manuscript document has been attributed to Joaquim Dâmaso (1777–1833), a disciple of Teodoro de Almeida. *A vida do Padre Teodoro de Almeida*, Lisbon,

to the design of John Theophilus Desaguliers – the curator of experiments of the Royal Society of London and an assistant to Isaac Newton.[13] A good acquaintance with Desaguliers's orrery may be assumed, as the descriptions of Teodoro's later planetariums (1796 and 1824), whether they are written by Teodoro or not, usually stress the fact that it surpasses that orrery by several features.

Volume 6 of Teodoro's *Recreação filosófica*, however, focusing on astronomy and presenting the heliocentric system, in very positive terms but still as an hypothesis among others, would be published later, in 1762, when Teodoro had already left Lisbon for Porto. It does not mention any demonstration apparatus. On the other hand, it becomes evident from the book that Teodoro had first-hand experience of astronomical observation. He was very interested in linking such observation, like other experiments, with modern explanatory hypothesis including heliocentrism and universal gravitation.

The copper plate print in volume 6 of *Recreação filosófica* represents the typical equipment of an astronomical observatory of the mid-eighteenth century that might reflect some of Teodoro's own memories (Illustration 13). Three gentlemen are collaborating on this scene: the observer sits on a terrace and looks through a telescopic sight of a quadrant or sextant. It was a common procedure at the time to check the clocks by altitude measurements of celestial bodies. The central figure stands in front of a large time piece reading out the seconds, and the third person perhaps records or checks the events in a book. The latter also contemplates the world system displayed on a large chart on the wall. The illustration convincingly conveys the idea that astronomical knowledge is rooted in a coordinated deployment of advanced apparatus, skilful sense perception, and rigorous documentation.

In the years up to the publication of this volume, Teodoro reportedly made astronomical observations in collaboration with – and possibly influenced by – his fellow Oratorian João-Baptista Chevalier (1722–1801). This activity was not carried out in isolation, but was seen by the authors as a contribution to a larger enterprise of the European Republic of Letters. The results of the observation of the lunar eclipse of 4 February 1757 were sent to the Royal Society of London.[14] Not surprisingly, Teodoro was elected fellow of the Royal

Arquivo Nacional da Torre do Tombo (ANTT), Manuscritos da Livraria, no. 2316. https://digitarq.arquivos.pt/ViewerForm.aspx?id=4248815 [Accessed 29 April 2024], microfilm no. 227.

13 Desaguliers seems to have developed his own version of orrery in collaboration with George Graham. See McEvoy, 'George Graham', pp. 235–50; Carpenter, *John Theophilus Desaguliers*, pp. 39–40.

14 Published in the *Philosophical Transactions*. Chevalier and de Almeida, 'Eclipsis Lunæ', pp. 376–77.

Illustration 13. Copper plate print by Miguel Rodrigues, 1762, from *Recreação filosófica*, 'Trata dos Ceos e do Mundo', vol. 6 (Lisbon 1762) (*Biblioteca Nacional de Portugal, Lisbon, TR. 6836 P*)

Society in London on 9 March 1758. Chevalier was also a correspondent of Joseph-Nicolas Delisle (1688–1768) in Paris, and his observations with Teodoro de Almeida reached Paris, integrating in this way the wide-reaching epistolary network of the French astronomer.

In 1769 Teodoro was driven out of Portugal for political reasons. There was not only his connection to figures of the Portuguese nobility, notably the Marquis of Távora (1723–1759), who challenged the centralising ambitions of King Dom José I (1714–1777) and his powerful minister, the future Marquis of Pombal (1699–1782). The minister enforced measures to restrain the activity and the teaching of the Oratorians in Lisbon. Several of his fellow Oratorians, like João Baptista and João-Baptista Chevalier, had been targeted by the minister's faction and threatened with imprisonment. This made Teodoro cautious.[15] In 1760 Teodoro moved first to the North of the country, to the city of Porto. From there he continued to publish his work on natural philosophy and finished the provisionally last volume 6. It was from Porto that Teodoro reported his observation of the rare transit of Venus on 6 June 1761.[16] The teaching by the Oratorian fathers in the city of

15 Contente Domingues, *Ilustração*, pp. 87–93; Malato Borralho, 'Teodoro de Almeida', p. 217.

16 Published later in: De Almeida, 'Observation', p. 352.

Porto was well developed and most probably included the demonstration of experiments using apparatus from a cabinet.[17] There is no information, however, whether Teodoro could have access to this instrument cabinet nor whether it included a planetarium. Finally, he crossed the border and fled through Spain to the French city of Bayonne in order to escape imprisonment and prosecution.

Outside Institutions: Exile Life in Bayonne (1769–1778)

According to the anonymous author of the *Life*, Teodoro's initial plan was to go to Amsterdam. If this is true, the choice of this city in a non-Catholic country might seem surprising. It would show that the Oratorian intended to rely mainly on his skills as presenter of physical demonstrations and philosophical experiments for a living.[18] On the other hand, Teodoro's former fellow observer, João-Baptista Chevalier, had found refuge in his exile in Brussels, in the Austrian Southern Netherlands, under the protection of Duke Louis-Engelbert d'Arenberg (1750–1820) before 1763.[19] Prospects for a new start in the Southern Netherlands were apparently not too bad for Oratorians, and Amsterdam may have been a station to reach that destination. Teodoro's voyage was interrupted, however, so that he never reached the Low Countries. Instead he settled in Bayonne on 25 February 1769 where he soon found a way to earn a living by private teaching.[20] This period is described in the *Life*:

> He moved into a house next the one of the Convent of the Visitation of Bayonne, and progressively occupied himself with the ministry of a confessor, dedicating himself to this in the days that were free from his teaching of experimental physics, and mathematics, by which he sought to make a decent living.[21]

Apparently, he did not stay in that French town in any specific capacity. In a letter to António Nunes Ribeiro Sanches (1699–1783), who was an exiled

17 Dos Santos, *O Oratório*, pp. 309–14.

18 This was noted by Contente Domingues, *Illustração*, p. 98.

19 De Ridder and Lemaire, 'Chevalier, Jean-Baptiste', cols. 180–81.

20 See MS *História da fundação do Mosteiro da Visitação*.

21 The 'couvents de la visitation' were a number of female cloisters whose sisters devoted themselves to charity work. 'Tomou caza cazualmente junto da Vizitação de Baiona; e pouco a pouco se foi empregando no Ministério do Confessionário, em que se occupava nos dias livres das suas lições de Fisica experimental, e Mathematica, com que procurava uma decente sustentação'. Cited after Santos, *Literatura*, p. 470.

Portuguese staying in Paris, Teodoro stressed his condition of a devotee of science without any formal titles, and with no need of them:

> I am not an academic, nor do I possess any honorary pedigree: I'm living in my home, secluded but known by those who seek me. The constant coming and going of persons spread the news about me more than I would like, and the Planetarium I made, and that I teach, and other objects that the audience likes, affords me the respect of all with no need for me to possess any title.[22]

From the correspondence of the two exiles we learn that Teodoro had up to twenty private students.[23] His teaching was experimental and very much appreciated. A letter from Ribeiro Sanches to Teodoro dated 26 September 1774 reveals some further aspects of the exiled priest's activities:

> I'm happy to see that you have decided to turn your thoughts to the pleasurable work of making thermometers and barometers, and that geographic map of wood (a new and noteworthy idea for teaching the blind); you're also interrupting sometimes your teaching activity, and that you want to make a *camara obscura*.[24]

One gathers that Teodoro considered his planetarium the centrepiece of his success. Here is what we know about this first version of Teodoro's planetarium: on the horizontal plane it represented the motion of the planets around the Sun with their proportional speeds, and the Earth's proper rotation. The planets including the Earth were driven – as was usual for this type of demonstration device – through a gear mechanism connected to a crank. We know about his initial conception of the planetarium thanks

22 'Eu não sou Académico, nem tenho grão algum onorífico: vivo em minha Caza, retirado e conhecido dos que me buscão, a passagem continua de gentes que vão e que vem me fazem mais conhecido do que eu quisera, e o Planetário que fis, logo que ensino, e outras coizas contentão o público fazem que me respeitem sem ter títulos de penachos'. Our translation from the citation in Silva, *A apropriação*, p. 93, who takes it from Azevedo, *P.e Teodoro de Almeida*, p. 314. These letters are preserved at the National Library in Vienna: Ribeiro Sanches, *Commercium epistolicum*, Vienna, National Library, ÖNB, Cod. 12713, Cod. 12714.

23 This is mentioned in a letter of 15 February 1774, according to Contente Domingues, *Ilustração*, p. 99.

24 'Alegro me que V[ossa] M[ercê] tomou a resolução de occupar o pensamento no trabalho divertido de fazer thermometros e barometros, e essa Carta Geografica em madeyra (idea nova e curiosa para ensinar um cego); VM tãobem le por pausas, e quer começar huã camara obscura'. Published in Rodrigues Gusmão, 'Um invento Portuguez', pp. 139–40.

to a series of manuscript notes of technical nature written in French and Portuguese that have survived.[25] This first planetarium is also known from later references.[26]

If we trust the later description, which points out the differences with regard to his 1796 planetarium, Teodoro's first mechanism from the 1770s implemented the following:

- the planets moved on clearly eccentric circles, making evident perihelion and aphelion (the closest and farthest distance from the Sun);
- the planets moved in their own plane and not in the ecliptic plane, making evident the nodes of the orbits. This made it possible to explain why the transits of Mercury and Venus happen only on rare occasions;
- the outer planets (Mars, Jupiter, and Saturn) were moved by the mechanism itself connected to the crank. These planets would not need manual setting. Yet, Jupiter and Saturn (the outermost planets, such as Uranus, had not been discovered yet) would make small steps once in a while, being at rest the remainder of the time;
- the Sun (shown as a sphere with its spots) had a proper rotational motion;
- the Earth-Sun system also showed the beginning and end of dawn;
- the Moon's illuminated side was always facing the Sun.

Justifiably or not, several of these features were considered improvements over Desaguliers's orrery (and also over other devices like the one built by Wright & Cole in 1747) where, for instance, the outer planets were placed manually. Moreover, and in particular, Teodoro claimed that the mechanisms (especially that of the Earth-Moon system) do not distract the attention of the audience.[27]

Somewhat connected to the planetarium project, Teodoro was also looking for a way to explain how the tides on both sides of the Earth were produced by the Moon's motion: on the side facing the Moon and the opposite one. Teodoro sought to buttress a theory of the tides that invoked, differently than Newton, the concept of centrifugal force exerted on the rotating Earth-Moon system. To show this, Teodoro conceived an additional mechanism.

Since 1774 Teodoro had speculated about the near end of Marquis Pombal's regime, and started considering his return to Portugal.[28] In anticipation of this he sought to sell the planetarium made in Bayonne together with other

25 Technical notes on a planetarium (in French), ANTT, Ms. da Livraria no. 1675 /2(1);(2);(3).

26 De Almeida, *Descripção*, p. 6.

27 'que desfigura o Planetario de Dezaguliers, e outros', De Almeida, *Descripção*, p. 13.

28 Contente Domingues, *Ilustração*, p. 100.

instruments at auction in 1775. The process of organising this auction had been tricky: he needed special authorisation from the Crown Procurator in Bordeaux. He was obliged to have a description of the apparatus printed (a description that he must have published, but that we have not yet been able to locate) that was then certified by municipal officials in Bayonne. This certification was delivered on 9 August 1775.[29] Teodoro must have finally managed to sell the planetarium as is attested also by another letter from Ribeiro Sanches, dated 18 January 1777:

> I would suppose that you received the benefit you hoped from the auction of the Planetarium: I'm happy that one of your students won the lot, and that it can be useful for the Sociedad Bascongada; because I think that one should begin the study of physics by astronomy [...].[30]

We can deduce that the planetarium went, for a good price, into the hands of someone who had attended Teodoro's lessons, and who had taken it to the Basque patriotic society. Teodoro became himself a member of the Real Sociedad Bascongada de los Amigos del Pais, in 1776. This regional academy, founded under the impulse of Count Peñaflorida (1729–1785), was based in the small Basque city of Vergara across the border in Spain. It acted as a patriotic society aimed at promoting local industry and education by institutional means, by publications, and public events. Teodoro appears in the published catalogue of members list of 1775–1776: 'Almeyda (el Padre Teodoro) profesor de matemáticas e fisica esperimental. L. en Bayona de Francia'.[31] The profile under which Teodoro is presented speaks to the way the Oratorians' devotion to experimental philosophy was perceived as the defining trait of his person.

It proves difficult to track further this first planetarium by Teodoro. Possibly, it circulated on and a later observer stated that it showed up in an 1824(?) article of 'Correio da Europa' (*Courrier de l'Europe*, a London based gazette written in French well circulated in Iberian circles), with no mention

29 See Municipal register of certification, Archives municipales de Bayonne, F.F. 602, Docs. 1774–1776, no. 86, no. 87; F.F. 554, no. 68, F.F. 509, Docs. 1675–1781, no. 18, transcribed in Caldas Franco Duarte, *A filosofia*, pp. 28–41.

30 'Estimarei que V[ossa] M[ercê] tirasse o lucro que esperava da lotteria do Planetario: fico contente que hum seu Discipulo sahisse premiado com elle, e que possa ser util á sociedade Vascongada; porque penso que se devia começar o estudo da Física pello da Astronomia: Mas vejo que em Castella ainda estão ligados com as correntes da Física dos Conventos, Thomistas, Scotistas, Reaes Nominaes e Integraes: Quando se acabará tanta parvoice fradesca?' Ribeiro Sanches, letter dated 18 January 1777, *Commercium epistolicum*. Vienna, National Library, ÖNB, Cod. 12713, Cod. 12714.

31 *Extractos de las juntas generales celebradas por la Real Sociedad Bascongada*, p. 96.

of Teodoro. The appropriation was then apparently rebutted by 'a friend' of Teodoro's in the *Jornal Literario*.[32] It was not unusual for the periodic press to discuss and present planetary mechanisms in the early decades of the nineteenth century in France (and elsewhere in Europe).[33]

As to many of Teodoro's other instruments, including a demonstration apparatus to show the effect of gravity producing the tides, they ended up in the (formerly Jesuit) college of the French city of Auch on the route from Bayonne to Toulouse.[34] This college was run by secular clergy, from the congregation of Notre-Dame de Garaison, and directed by the clergyman Marignan. It seems that Teodoro was once even offered to supervise the secondary school teaching at Auch, which he declined. Nevertheless, it is attested that Teodoro gave a short course of experimental philosophy there before handing over his collection in February 1777.[35]

The exile of Teodoro de Almeida in Bayonne constitutes a revealing episode with regard to his dedication to the sciences. True to his calling, the Oratorian never ceased his activity as confessor and spiritual mentor, which remained the central motive of his life. At the same time, uprooted from the congregation of his country, his living would depend on the earnings from giving private courses on geometry and physics. Teodoro appears to have entered easily into the new surroundings of the French city. He clearly enjoyed teaching and creating instruments for teaching, as becomes clear from the list of his creative developments. As these attested to Teodoro's knowledge, imagination, and craftsmanship, he became highly valued in the context of the coastal town. Beyond attracting visitors and students, Teodoro was also offered jobs as a school director in Auch (which he refused) and he was invited to act as an examiner of aspiring hydrographers in Bayonne. Further proof is provided by the number of interested people who sought to meet Teodoro and see him perform demonstrations and give explanations on physics, geography, and astronomy. One might suspect that his visibility as a devotee of science was even more determinant in defining his position in society than his quality as a clergyman, as can be judged from the way his membership admission to the Basque society was registered.

The planetarium stands out as the keystone of Teodoro's achievements during the Bayonne period. Not only did it constitute the most complex of

32 We were unable to locate either of these publications. See: *Life* § 75.

33 For example, in the *Bazar parisien* and *Annales maritimes et coloniales*.

34 See Silva, *A apropriação*, p. 93. The collection was bought in 1777 for 1800 lires. See Caldas Franco Duarte, *A filosofia*, p. 41, who cites: Bordes, *D'Etigny*, vol. 2, p. 894, n. 84.

35 Contente Domingues, *Ilustração*, pp. 100–101.

all devices he had developed, but it would also have brought him the largest monetary profit when he sold it at auction. Beforehand, a description of the device would be officially certified by experts on behalf of the municipality of Bayonne, a recognition that scientific apparatus rarely obtains in such bureaucratic moulds. It indicates that Teodoro's activity as cultivator of the sciences was more than a private affair, but that the official representatives of the city recognised its importance.

Return to Lisbon: 1778

Teodoro was finally able to move back to Portugal when King Dom José died and the regime in Portugal changed in 1777. From this moment on he enjoyed the confidence of the highest-ranking nobles at court, including influential women.[36] With his return Teodoro's situation changed dramatically, which also enabled him to pursue his devotion to science teaching on a new level. On the other hand, coming back into the closely-knit network of the Oratorian congregation of Lisbon, courtly, and also academic circles would require much attention and possibly leave less time for personal endeavours. Teodoro would focus on participating in the foundation of the Academia Real das Ciências (Academy of Sciences of Lisbon) that rapidly materialised in 1779, and lobbying for the foundation of a religious institution for young women, the monastery of the order of visitation (achieved in 1784).

Nevertheless, once back in Lisbon, Teodoro soon started to make a second planetarium. He seems to have begun this project quickly as he must have felt the lack of an adequate device for his usual lectures on the 'Copernican' system. He could use the plans he still had in his notes from Bayonne. Moreover, this time he was in the comfortable position of recurring to the workshop of the Arsenal Real do Exército, the army arsenal and its workshops, to have it built. Manuscript notes preserved today show that he was directly involved in buying some of the spare pieces at the foundry.[37]

Teodoro appears to have produced, at the beginning, a planetarium in the moulds of the one he had conceived in Bayonne: a disposition in a horizontal plane, without including Uranus that was still unknown and identified as a planet only in 1781. Significantly, the Oratorian also came up with some improvements: he made the motions of Jupiter and Saturn continuous (instead of step-wise) and, in addition, made one turn of the crank correspond to one day (instead of the previous, more or less arbitrary, four days and a seventh). On the whole, this second planetarium was an

36 *História da fundação do Mosteiro da Visitação em Lisboa*, ANTT, Ms. Livraria, no. 661.

37 Technical notes on a planetarium, ANTT, Ms. Livraria no. 1675 (2), I & II.

improvement, but still fairly simple, and probably not very different from the one built in Bayonne.

The literature dealing with Teodoro's activity as a populariser of science has never specifically focused on his achievements as a planetarium builder. Therefore, it is not surprising that it has not been pointed out that the Oratorian did not just build two but actually three planetariums. Teodoro had offered his second planetarium to the ruling prince Dom João. The documents of this exchange make it clear that the prince, instead of simply accepting the offer, in turn provided the funds to have a third planetarium constructed. The explicit motive for doing so was to avoid that Teodoro would lack an instrument for his teaching at the House of the Oratorian Congregation, the Casa do Espírito Santo in Lisbon.

These funds were sufficiently generous for Teodoro to think of several improvements. This process resulted in the last and most developed version of Teodoro's planetariums, one that was much more ornate and incorporated additional parts and ideas. Its construction was completed in the year 1796.[38] To figure out the details of this planetarium, four different types of historical sources are today at our disposal, as we will see: a printed pamphlet with a detailed explanation of the parts (in two editions 1796 and 1797), a lithographic depiction (1824), a document from the state administration (1843), and even a wooden fragment of the original planetarium preserved to this day.

It turns out that, also visually and structurally, this planetarium differed radically from the previous ones. The lithograph, produced a few decades after the instrument was constructed, makes these differences most apparent (Illustration 14). First and foremost, the planetarium showed the ecliptic plane in a vertical position, like a clock face, with the explicit intention of making it better adapted for demonstration in front of a numerous audience.[39] On the lower left and right hand sides, it showed two sculptures allegorically representing Astronomy (Urania) and Time (Chronos). Urania's hand pointed to a circular sign inscribed with each planet's period of revolution around the Sun. Chronos, for his part, pointed towards a circular dial displaying the lapse of time in months and days as one turned the crank.[40] This disposition is strongly reminiscent of a copper plate

38 *Life*, § 75.

39 Other devotees also found the means to adapt their demonstration devices to be more efficient in their specific situations, as in the case of Eise Eisinga who built his planetarium into the ceiling of his living room. See the chapters Arjen Dijkstra and Fokko Jan Dijksterhuis in this volume.

40 De Almeida, *Descripção*, p. 4.

print representing the first verse of *Genesis* drafted by the famous painter and engraver Gerard Hoet (1648–1733) from The Hague. That print shows a 'Copernican' disposition of the heavens, including the moons of Jupiter, that looks very similar to an orrery, and a sectional representation of the Earth joined on an altar-like vertical structure.[41] The altar includes quite similar allegorical figures like those of Teodoro's planetarium equipped with the same attributes: Eternity as Astronomy is associated with an armillary sphere and Time or Chronos with a scythe.[42] Hoet's image was distributed with a series of engravings representing scenes from the biblical Old and New Testament printed, published, and sold by Pieter de Hondt (1696–1764), in 1728.[43] It is not impossible that it served as a model for the artists assisting Teodoro in Lisbon. This ornate and more perfect planetarium by Teodoro was clearly intended to fit its dedicatee, Prince Dom João, such as the dedicatory inscription on the top of the instrument indicated. At the same time, the framing of the device as an allegory of God's initial act of creation clearly situated astronomical learning within religious devotion and put Teodoro's dedication to scientific learning into an appropriate perspective.

From the later administrative document we learn that this same planetarium also had a drawer containing a tellurium-type mechanism and a device for demonstrating the tides:

> a small instrument that is kept in the drawer of the Machine demonstrates the principles of days, nights, dawn and dusk, the duration of days in different climates and seasons of the year. The Moon as the satellite of the Earth is seen rotating around it, producing its periodic and synodic revolutions; showing its various aspects, and completing its year around the Sun, and uncovering to the eyes the reason of the Computus of the Epacts; the points of intersections of its orbit with the ecliptic are shown; how, when, and why the solar and lunar eclipses happen; and the cause of the tides, and of those termed high tides, to which effect a second instrument is kept in the same drawer.[44]

41 See the chapter 'Imagining Our Place amongst the Stars' by Wouter de Vries in this volume.

42 Explanation of these figures is given in De Hondt's Dutch edition of *Taferelen*, p. 1.

43 De Hondt, *Taferelen*. The plates were also published with a frontispiece in French, in the same year: De Hondt, *Figures de la Bible*.

44 'hum pequeno instrumento que se conserva na gaveta da Machina mostra a razão dos dias, noites, e crepúsculos, e da grandez dos dias dos diversos climas e estações do anno. A Lua como Satelite da Terra se vê girar a volta della, fazendo as suas revoluções periódicas e synódicas; mostrando os seus diversos aspectos, completando o seu anno à roda do Sol, e descobrindo visivelmente a razão do computo das Epactas; mostrão-se os pontos da intersecção da sua órbita

From the year 1796, when the planetarium was constructed, a printed pamphlet about the mechanism survives giving further details about Teodoro's intentions and also providing an impression of how the different parts were explained in public.[45] The booklet was issued at the Royal Press (Regia Officina Typografica) in Lisbon, showing that the Oratorian received full backing from the Crown for his enterprise. Curiously, only in a second print run in 1797, with slightly altered pagination, was a page on the planet Uranus added. Uranus was clearly included in this third version of the planetarium. This can be ascertained because we possess a visual depiction of this 'vertical' planetarium. A lithographic picture was issued in 1824, representing the new planetarium in detail. The planets are labelled by the initial letter of their names, for instance, 'M' for Mars, 'J' for Jupiter, 'S' for Saturn, and 'U' for Uranus. They appear as small spheres, illuminated on the side turned towards the Sun, being moved along the slots of their orbit by an invisible mechanism.

The lithograph also shows the dedicatory inscription 'Joanni Principi – Theodorus de Almeida ex Oratorio O' (For prince Dom João – Teodoro de Almeida of the Oratorian Congregation o[rdered it?]), as well as the place and date of its making ('Arçenal Real do Exército, Lisboa, 1796'). The lithograph's caption reads:

> Planetarium, invented in France by Father Teodoro d'Almeida, and by the same enhanced in Lisbon, which is shown at the House of the Holy Spirit of the Congregation of the Oratory. Represented on stone by Caetano, 1824.[46]

Despite its dedication to Prince João, it seems that the third planetarium continued to be used for teaching at Casa do Espírito Santo, the residence of the Oratorian congregation. This is what research by the first author and

com a eclíptica; como quando e porque razão acontecem os Eclipses de Sol e da Lua; e a razão das marés, e daquelas a que se da o nome de grandes marés, para o que serve outro instrumento que se conserva na mesma gaveta'. Transcribed from 'Representação de Vicente Ferreira de Sousa Brandão, Encarregado do Real gabinete de Physica d'Ajuda, requisitando um Planetário e um Pluviómetro, que pertencem a Sua majestade e se acham extraviados; bem como todos os mais pertences que dizem respeito a este negócio. 26 de Março de 1843', ANTT, Casa Real, Caixa 4003.

45 De Almeida, *Descripção.*

46 'Planetário, inventado em França pelo P. Theodoro d'Almeida, e pelo mesmo augmentado em Lisboa e se mostra na Caza do Espirito Santo da Congegraçam do Oratorio. CAETANO dezenhou em pedra em 1824'. Several surviving copies of this large-sized print (35.5 × 16 cm) have been located: one in the *Centro Documental da Marinha*, Lisbon; another five copies at the National Library of Lisbon, including BNP S.A. 1448//3 V.; and four copies BNP, Iconografia E. 1126 V., E. 1507 V., E. 2012 V., E. 2226 V.

Illustration 14. Teodoro de Almeida's orrery. Lithograph drafted by Caetano Aires de Andrade, 1824. (*Collection Biblioteca Nacional de Portugal, Lisbon, E. 1126 V*)

colleagues could establish so far.[47] These historians have also identified the subsequent whereabouts of this representative piece: from the Casa do Espírito Santo, where it was at least until 1824, it moved to the Biblioteca Pública (located in the former Convento de São Francisco) in 1836 (all male religious orders were abolished in 1834). Sometime later, in 1842, its presence is attested at the recently founded (1837) Escola polytécnica that occupied the premises of the former Jesuit novice school. Finally, the planetarium left the main building of the Escola polytécnica and remained in the astronomical 'teaching observatory' located in the gardens in the rear of the building. This was the place where we, nearly two hundred years later, rediscovered a fragment of the planetarium's ornate base, the only material remains of Teodoro's planetarium that seem to have withstood the ravages of time.[48]

Conclusion

The biography of Teodoro de Almeida constitutes an interesting example of the changing roles, and the variety of motives for using and/or producing planetariums and other devices of experimental philosophy during the eighteenth century. While Teodoro's position and social context underwent dramatic changes throughout his career, his ability to convey astronomical and physical knowledge through live demonstration using all kinds of apparatus would serve him differently at each stage. Significantly, one may observe that these differences in social configuration are also reflected in the material structure and design of the three different planetariums built under Teodoro's supervision.

At the beginning of his trajectory in Lisbon, Teodoro was part of a well-established system: he organically grew into his role of assistant professor and then principal teacher of natural philosophy at the well-regarded convent-palace of the Oratorians. The House, through the energetic initiative of João Baptista, and generously granted royal support, set up a well-equipped teaching cabinet. Teodoro could use the orrery of the cabinet, quite probably one designed by Desaguliers, to expound the heliocentric system. There was no need to start building one himself. Only in Porto, his first station

47 The colleagues include David Felismino, Ricardo Barbosa, and António Perestrelo de Matos. This research was carried out as part of the project 'On the Instruments' Trail: Exploring Royal Cabinets of Natural Philosophy in Portugal (18th–19th Century)', supported by the Portuguese Foundation for Science and Technology (ref. FCT/MCTES, PTDC/HIS-HCT/098970/2008).

48 This fragment of the planetarium is now preserved at the Museu Nacional de História Natural e Ciências, MUHNAC, inv. no. MUHNAC-UL, UL3943.

of exile, he would come to feel the lack of a demonstration apparatus in the shape of an orrery.

During his exile in Bayonne, France, the situation was rather different. Teodoro was alone, in the sense that he could not count on a local Oratorian institution. Beside his work as a confessor he needed to develop his physics teaching for a living, which required that he build instruments by himself. We have no information about the funding and the process of instrument making, but we know that Teodoro succeeded in producing an orrery-type mechanism himself. It would be similar to that of Desaguliers but avoiding the aspects that Teodoro considered serious flaws and improving it.

There is no doubt that Teodoro was an excellent communicator and also that he sought to adopt most innovative mechanical means to visualise scientific content. Both these aspects earned Teodoro a good reputation as a teacher of natural philosophy and allowed him to make a living. We saw that Teodoro's status hinged on this success. Nevertheless, Teodoro apparently never took his engagement in science and teaching as an end in itself. This becomes clear when the opportunity appeared, and he did not hesitate to sell his instrument collection, including his planetarium, to generate additional income.

A fourth stage in life begins when Teodoro is back in Lisbon. He immediately achieved a well-respected status, became a confessor for high-ranking personalities, and the driving force behind the foundation of religious institutions. Although Teodoro seemed to have intended the making of a new planetarium as early as 1778, we do not know when he finished it completely. This time he would have the help of the artisans of the workshop of the army arsenal. We may reasonably assume that Teodoro would use it once again in lessons for students of the Oratorian House of Espírito Santo. Teodoro remained true to his love of natural philosophy and experimentalism, but his profile as a devotee of science had no part in the newly found reputation in Lisbon. His rise to influence relied more on his reputation of a loyal clergy, confessor, and counsellor who had maintained relationships with several high-ranking personalities over many years.

It appears that the successive planetariums used and constructed by Teodoro de Almeida were increasingly sophisticated in that they included more and more subtleties of nature. This observation is in striking agreement with the changing audience of Teodoro. Each station of his life led him to teach persons of ever higher status and better education. There is no direct causality involved in this mirroring of the social context in the planetarium's materialisation, it simply shows to some extent that design decisions are motivated not only by the natural knowledge available but include the anticipation of the audience's point of view and its expectations.

It would be an exaggeration to think that Teodoro dedicated himself nearly exclusively to his calling as a teacher of natural philosophy and designer of planetariums and instruments. It is true, however, that his nominal function as a confessor and member of the Oratorian congregation became secondary during a portion of his life. One may surmise that Teodoro fulfilled what Oratorian clerics aimed at, namely using his specific talents to spread religious faith and evangelise those he reached, in his case, by expounding the orderly complexity of God's creation as it is apparent in the natural world.

Teodoro de Almeida was far from an exception as a devotee of science among European catholic clergymen of the eighteenth century, be they Oratorian, Jesuit, or of any other order. He certainly stood out by his extraordinary talent as a communicator to a non-scholarly public, and a talented conceiver of demonstration instruments, including planetariums. His contribution to the cultivation of science locally, in Lisbon and in Portugal more generally, is clear. By way of his planetarium Teodoro left a legacy of teaching astronomy using an orrery. Teodoro's activity, even after his death, was proudly celebrated, as the later lithograph edition attests, and inspired the subsequent, more institutionalised, uses of his planetarium in the context of the government-founded Escola polytécnica.

Bibliography

A vida do Padre Teodoro de Almeida. Lisbon, Arquivo Nacional da Torre do Tombo (ANTT), Manuscritos da Livraria, no. 2316. https://digitarq.arquivos.pt/ViewerForm.aspx?id=4248815, microfilm no. 227 [accessed 29 April 2024].

Annales maritimes et coloniales (1820). https://books.google.pt/books?id=-aRAAAAAcAAJ&pg=PA444 [accessed 29 April 2024].

Azevedo, M.L., *P[adr]e Teodoro de Almeida, subsídios para o estudo da sua vida e obra* [Dissertation] (Coimbra 1959).

Bazar parisien, ou Annuaire raisonné de l'industrie des premiers artistes et fabricans de Paris (1826). https://gallica.bnf.fr/ark:/12148/bpt6k65838751/f96.item [accessed 29 April 2024].

Bordes, M., *D'Etigny et l'administration de l'entendance d'Auch (1751–1767)*, vols. 1–2 (Auch 1957).

Caldas Franco Duarte, M.T., *A filosofia do padre Teodoro de Almeida, através da 'Recreação Filosófica'* [Dissertation] (Lisbon 1959).

Carpenter, A.T., *John Theophilus Desaguliers: A Natural Philosopher, Engineer and Freemason in Newtonian England* (London 2011).
Chevalier, J.B., and T. de Almeida, 'Eclipsis Lunæ die 4a Februarii, Ann. 1757 habita Ulissipone à Joanne Chevalier presbytero congregationis oratorii, Regiæ Londinensis Societatis socio, regiæque scientiarum Parisiensis Academiæ correspondente, et a Theodoro de Almeida ejusdem congregrationis presbytero, ac physicæ publico Professore', *Philosophical Transactions* 50 (1758), pp. 376–77.
Contente Domingues, F., 'Um Projecto Enciclopédico e Pedagógico: A Recreação Filosófica de Teodoro de Almeida', *Revista da História das Ideias* 10 (1988), pp. 235–48.
Contente Domingues, F., *Ilustração e catolicismo: Teodoro de Almeida* (Lisbon 1994).
De Almeida, T., 'Observation du passage de Vénus sur le disque du Soleil, faite à la ville de Porto en Portugal, en l'année 1761', *Mémoires de Mathématique et de Physique, Présentés à l'Académie Royale des Sciences, Paris* 6 (1774), p. 352.
De Almeida, T., *Descripção do novo planetário* (Lisbon 1797).
De Hondt, P., *Figures de la Bible* (The Hague 1728).
De Hondt, P., *Taferelen der Voornaamste Geschiedenissen van het Oude en Nieuwe Testament* [...] (The Hague 1728).
De Ridder, A., and C. Lemaire, 'Chevalier, Jean-Baptiste', in: *Biographie nationale* 43, suppl. 15 (Brussels 1983), cols. 180–97.
Dos Santos, E., *O Oratório no norte de Portugal: Contributo para o estudo da história religiosa e social* (Porto 1982).
Dos Santos, E., 'Oratorianos', in: J.E. Franco et al., ed., *Dicionário Histórico das Ordens e Instituições afins em Portugal* (Lisbon 2010), pp. 231–40.
Extractos de las juntas generales celebradas por la Real Sociedad Bascongada de los Amigos del Pais en la villa de Vergara (Vitoria 1776).
Feijoo, B.J., *Teatro critico universal* (Madrid 1726–1739).
Findlen, P., 'Science as a Career in Enlightenment Italy: The Strategies of Laura Bassi', *Isis* 84, no. 3 (1993), pp. 441–69.
Fiolhais, C., 'Os diálogos filosóficos do padre Teodoro de Almeida', *Limite* 11, no. 1 (2017), pp. 89–110.
Hellyer, M., *Catholic Physics: Jesuit Natural Philosophy in Early Modern Germany* (Notre Dame 2005).
História da fundação do Mosteiro da Visitação em Lisboa, Lisbon, Arquivo Nacional da Torre do Tombo (ANTT), Manuscritos da Livraria, no. 661.
Leone, M. et al., 'La Fisica nei "Gabinetti di Fisica" dell'Ottocento: Il caso dell'Università di Genova', *Giornale di fisica* 50, no. 3 (2009), pp. 167–84.
Malato Borralho, M.L., 'Teodoro de Almeida. Entre as histórias da História e da Literatura', in: L.A. de Oliveira Ramos et al., ed., *Estudos em homenagem a João Francisco Marques* (Porto 2001), pp. 213–27.

McEvoy, R., 'George Graham and the Orrery', *Nuncius* 35, no. 2 (2020), pp. 235–50.
Municipal register of certification, Bayonne, Archives municipales de Bayonne, F.F. 602, Docs. 1774–1776, no. 86, no. 87; F.F. 554, no. 68, F.F. 509, Docs. 1675–1781, no. 18.
Pereira Gomes SJ, J., 'João Baptista e os peripatéticos', *Brotéria* 39, no. 2 (1944), pp. 121–37. Re-edited in: H. Leitão and J.E. Franco, ed., *Jesuítas, Ciência e Cultura no Portugal Moderno: Obra Selecta do Pe. João Pereira Gomes SJ* (Lisbon 2012).
'Representação de Vicente Ferreira de Sousa Brandão, Encarregado do Real gabinete de Physica d'Ajuda, requisitando um Planetário e um Pluviómetro, que pertencem a Sua majestade e se acham extraviados; bem como todos os mais pertences que dizem respeito a este negócio. 26 de Março de 1843', Arquivo Nacional da Torre do Tombo (ANTT), Casa Real, Caixa 4003.
Ribeiro Sanches, A., *Commercium epistolicum*, Vienna, National Library, ÖNB, Cod. 12713, Cod. 12714.
Rodrigues Gusmão, F.A., 'Um invento Portuguez', *Archivo Pittoresco* 11, no. 8 (1868), pp. 139–40.
Santos, Z., *Literatura e espiritualidade na obra de Teodoro de Almeida: 1722–1804* (Lisbon 2007).
Schaffer, S., 'Natural Philosophy and the Public Spectacle in the Eighteenth Century', *History of Science* 21 (1983), pp. 1–43.
Silva, J.A., *A apropriação da filosofia natural em Teodoro de Almeida (1722–1804)* (Lisbon 2009).
Silva, J.A., 'Recreação Filosófica, an Eclectic War Machine', in: L. Ferrão and L.M. Bernardo, ed., *Views on Eighteenth Century Culture: Design Books and Ideas* (Newcastle upon Tyne 2015), pp. 206–18.
Taton, R., ed., *Enseignement et diffusion des sciences en France au 18e siècle* (Paris 1986).
Technical notes on a planetarium (in French). Arquivo Nacional da Torre do Tombo (ANTT), Manuscritos da Livraria, no. 1675 /2(1),(2),(3).
Tosca, T.V., *Compendio Mathematico* (Valencia 1707–1715).
Wyka, E., 'Collections of Experimental Natural Philosophy in Eighteenth-Century Poland', in: J. Bennet and S. Talas, ed., *Cabinets of Experimental Philosophy in Eighteenth-Century Europe* (Leiden/Boston 2013), pp. 173–93.

About the Authors

Luís Tirapicos is a researcher at the Interuniversity Center for the History of Science and Technology at the University of Lisbon.

Samuel Gessner is an associate researcher at the Department of History and Philosophy of Sciences at the University of Lisbon.

9. Imagining Our Place amongst the Stars: Visual Astro-knowledge and Early Modern Understandings of the Universe

Wouter de Vries

Abstract: Between 1600 and 1800, ideas of what space *is* changed fundamentally: from a set of solid spheres, space became an empty space, in which objects move or are moved. The published works reflect a much broader interest in what goes on above our heads – gazing at the stars became a favourite pastime. Artefacts and publications reflect and shaped an understanding of space as a space, and the imagination of the heavens is central to their constructions. This chapter describes the context within which many devotees of science worked and traces how prints formed an important element in the movement of knowledge between different spheres.

Keywords: cultural history, knowledge production, astronomy, printed images

This chapter investigates early modern understandings of the universe. From the seventeenth through the eighteenth century, contemporary ideas on the universe shifted from a notion of eternal spheres to that of a spatial extension open for exploration. The echoes of the 'battle of systems', in which Kepler's, Brahe's, Copernicus's, and other notions of the universe were at stake, were still being felt and they touched upon more than knowing which body stood at the centre of the universe. The debates found expression in print – text, diagram, image, and even frontispiece – and as the eighteenth century moved on, engaged a much wider audience than just a handful of scholars and stargazers. What knowledge of the universe did people have?

Bijleveld, Nikolaj, Arjen Dijkstra, and Samuel Gessner, eds. *Cultivating and Communicating Natural and Technical Knowledge around 1800: Devotees of Science.*
Taylor & Francis Group, 2025
DOI 10.5117/9789463728577_CH09

What understanding of it lay behind their fear of a conjunction? And most importantly, what aspects does this knowledge consist of – was it purely informational in nature, or did it also entail a certain sense of what the universe is – and our place in it too? By tracing the communication of different aspects of knowledge through printed images, this chapter explores the idea that understanding the universe was just as much a matter of the imagination as it is of knowing.

The eighteenth century saw a surge in publications that 'popularised', to a certain extent, knowledge about the heavens. Through celestial atlases, encyclopaedias, and dictionaries, but also through novels and prints, a non-scholarly audience was provided with an increasing number of means of informing themselves about what went on above their heads. Striking about many of these publications is the amount and quality of visual material they contain. Celestial atlases such as Doppelmayr's *Atlas coelestis* (1742)[1] or (biblical) encyclopaedias such as Johan Scheuchzer's *Physica sacra* (1731–1735) contain many page-filling, skilfully produced engravings. However, less ambitious projects such as Eberhard Kindermann's *Vollständige Astronomie* (1744) also relied heavily on visual material. Importantly, this was not a paper-only affair: public gatherings were organised in which scholars and amateurs alike gazed at the heavens, while planetariums or orreries gained popularity and became part of a wider, popular culture of science.[2]

The knowledge that is communicated in these publications is complex and consists of different aspects. On the surface, knowledge about the heavens seems to be mostly a question of knowing how the universe 'works'. Does the Earth revolve around the Sun, or the Sun around the Earth? How many moons does Jupiter have? How far are all the planets removed from each other? This knowledge is communicated through a medium, and with that comes a certain episteme. It matters whether it is communicated through image or text, through planetarium or print. Seeing a diagram of the solar system is entirely different from seeing an allegory of the seven planets, not only because it tells a different story about that system, but also because it tells the story differently. Finally, I would argue that there is a third aspect that plays a role: the cultural framework in which this knowledge gains meaning. What *is* this cosmos, and how does one imagine it – as an empty vacuum or a series of solid spheres, a macrocosm of cosmic forces or a place which harbours alien life? This third aspect of knowledge in particular is

1 Several of the plates in this atlas had been published earlier by Johann Baptist Homann, for example in his *Neuer Atlas*. Van Gent, 'The *Atlas Coelestis*'.

2 Zuidervaart, *Van 'Konstgenoten'*, pp. 43–98. See also Genuth, *Comets*, pp. 3–13.

not only dependent on scholarly developments, but is intertwined with a broader, cultural imagination.

In this chapter, I explore the changing conception of the heavens in the late seventeenth and eighteenth centuries through the lens of these different aspects of knowledge. In particular, I will investigate how knowledge of the heavens is disseminated through printed images. My argument will be that through visual illustrations in astronomical works – particularly those that are not at the heart of the scholarly enterprise – we can trace a conquest of the heavens that represents it as a spatial, visible, and explorable phenomenon. Understanding how the heavens were known includes *how* they were known: space is something that needs to be imagined before it can be investigated, and its investigation changes our imagination of it.[3] While the approach of this contribution is of a more general nature than that of some of the other chapters, we must bear in mind that worldviews such as addressed here inform and are shaped by the work of devotees as investigated in this volume.

Over the past few decades, much work has already been done on the study of early modern astronomical imagery. We now know much more about the epistemic role of astronomical diagrams, their relation to observation, and the production of these illustrations as a collaboration between scholar and artisan.[4] Historians have also problematised printed images as a medium. For example, Crowther and Barker have suggested in their 'Training the Intelligent Eye', that illustrations in sixteenth- and seventeenth-century astronomical textbooks served not only to teach the young pupil about the movement of the heavens; they were also intended to teach and instil a particular way of *looking* at the universe – to train an intelligent eye. Crowther and Barker's reading (or viewing) of this material reflects a multi-layered approach to epistemic images: one image can, at the same time, fulfil several epistemic functions.[5] They also, rightfully, quote Lüthy and Smets who remind us that every illustration or printed image is 'historically contingent'.[6] Finally, some (art) historians have focused on less-scholarly

3 This idea has been explored by Aït-Touati in *Fictions*. See also Theisohn, Buzek and Cwik, 'Astronomie und Weltraumfiktion', pp. iv–x.

4 Müller, 'Telescopic Observation', pp. 355–79. I also refer to the special edition of the *JHA* as a whole, as well as part two: *Journal for the History of Astronomy* 42, no. 1 (2011). Reeves, *Painting the Heavens*. Scientific illustrations as a whole have been studied extensively over the past few decades. See for example Lefèvre et al., *The Power of Images*, or Kusukawa and Maclean, *Transmitting Knowledge*.

5 Crowther and Barker, 'Training the Intelligent Eye', pp. 429–70.

6 The article they refer to is: Lüthy and Smets, 'Words', pp. 433, 437.

publications, or connected astronomical images to a broader sense of the world or the universe.[7] Lucia Ayala, for example, stresses the need for looking at visualisations of the different theories of the existence of other solar systems in the universe as an essential element of early modern cosmology.[8] Ingrid Rowland, in turn, stresses that 'the history of [ideas] is not only a history of science; it is always, equally a history of the humanities'.[9]

I intend to build on previous work by further exploring astronomical illustrations as a source for reconstructing the eighteenth century 'sense of space', while also problematising their role as epistemic images. I will investigate whether astronomical illustrations from the seventeenth and eighteenth centuries, even frontispieces and maps in celestial atlases, do not only teach the viewer knowledge of the orbits of Mars and the face of the Moon, but also serve a more rhetorical function, to 'frame' this knowledge in such a way that it helps the modern reader trace conceptions about what the universe was thought to be.

Scholarly Perspectives

Astronomy as a science developed significantly during the Renaissance. Two developments gaining prominence in the seventeenth century stand out in the context of this chapter: the spatialisation of space and the observation and visualisation of heavenly bodies. In the middle of the twentieth century, Alexander Koyré's *From the Closed World to the Infinite Universe* (1957) concluded that at the heart of the Scientific Revolution lay a 'geometrization of space', which turned the universe from a solid set of spheres into an infinite space in which the focus lay on geometrical relations.[10] While Koyré's book has been extensively revisited, and his theory has been debunked on some points, the idea that space spatialised over the course of this period still stands strong. W. Randles, in *The Unmaking of the Christian Cosmos 1500–1760* (with the apt subtitle 'From Solid Heavens to Boundless Aether') carefully studies the adoption of this perspective throughout the learned world – concluding that this was a rather lengthy process that stretched well into the eighteenth century.[11]

7 See for example Cosgrove, *Apollo's Eye*.

8 Ayala, 'The Universe in Images', pp. 283–303.

9 Rowland, 'Representing the World', pp. 75–105.

10 Grant, *Much Ado*, p. 233.

11 'It was only after the middle of the eighteenth century, with the increasing regard for the Newtonian system and with the liberalization of the intellectual climate in Rome [...] [that resulted in] bringing to an end the medieval cosmos'. Randles, *The Unmaking*, pp. 217–18.

The religious and theological implications of the spatialisation of space were vast.[12] Koyré suggested that this geometrisation of space, as he calls it, was the end for the idea that the universe was organised according to a metaphysical order. In the words of one of his later critics, 'in infinite space there was no longer any set of absolute coordinates with which to determine what was physically superior to what'.[13] More recently, the idea that the spatialisation meant the end of a metaphysical understanding of the cosmos has been nuanced. Quoting Gottfried Wilhelm Leibniz (1646–1716), Antognazza suggested that what happened is in fact a divergence of the physical and metaphysical rather than a triumph of one over the other.[14] Nevertheless, the end of the idea that the universe was limited, the stars were fixed, and the Earth was the centre of it all had to mean the 'unmaking of the Christian cosmos'.[15]

With space the subject of natural philosophy as well as astronomy, scholars increasingly framed space as the playground of physical forces and processes. René Descartes (1596–1650) famously imagined space as a vast 'sea' of particles, in which the planets and other heavenly bodies were moved by vortices within that liquid.[16] Isaac Newton's (1643–1727) theory of gravity similarly provided an explanation for the movement of planets in a universe that was no longer a series of concentric spaces.[17] The space that resulted from these theories was a place in which all sorts of things could happen, and from which all sorts of phenomena could be explained. For example, William Whiston (1667–1752), the successor of Newton on the chair of Mathematics in Cambridge, hypothesised that the Deluge was the result of Earth nearly colliding with a comet, the watery tail of which flooded the planet – indeed, space was still God's playground.[18]

Reimagining the universe as an open, infinite space also provided the opportunity to imagine the existence of worlds beyond our own. As early as the sixteenth century, the Italian philosopher Giordano Bruno (1548–1600) was one of the first to propose a theory of the 'plurality of worlds', as the

12 See also, for example, Grant, *Much Ado*, p. 241: 'During the seventeenth century, those who attempted to describe the properties of space usually found it essential to consider as well its ontological relations to the deity'.

13 Antognazza, 'Leibniz', p. 311.

14 Antognazza, 'Leibniz', pp. 312–13.

15 I refer again to the title of Randles, *The Unmaking*.

16 On Descartes's ideas of space, see Gaukroger, *The Emergence*, pp. 304–17.

17 Newton's conception of space is, of course, much more complex than this. For an overview, see DiSalle, 'Newton's Philosophical Analysis', pp. 34–60.

18 See Whiston, *A New Theory of the Earth*. On Whiston's work in general, see Force, *William Whiston*.

idea had become known by the seventeenth century.[19] Over the course of the seventeenth and eighteenth centuries, the idea would gain large traction both amongst scholars and in the cultural imagination of Europe, to the point where it went 'from being a belief of a few to a dogma taught in scientific textbooks and preached from pulpits'.[20] Examples of plurality thesis are offered by many scholars – in his 1698 *Cosmotheoros*, Christiaan Huygens (1629–1695) reiterates the idea of the existence of extra-terrestrial life, but the best-known proponent of the theory is of course Bernard de Fontenelle, whose *Entretiens sur la pluralité des mondes* (1686) was reprinted several times and reached a wide audience.[21]

In addition to this process of spatialisation, the seventeenth century witnessed a change (or innovation) in the methods through which scholars came to understand the heavens. Aided by instruments, scholars increasingly sought a more, or different kind of ocular engagement with the different bodies in the universe. The development of increasingly powerful telescopes offered not only a way of tracing the movement of heavenly bodies, but also of making observations on their appearance. In the early stages of this development, according to Albert van Helden, 'the telescope [...] produced information that could not always be easily encapsulated in the time-honoured media of astronomy: words, numbers and diagrams'.[22] The efforts of astronomers such as Galileo, Cassini, and Hevelius resulted in visualisations of heavenly bodies, that in some cases were 'valid for almost 150 years'.[23] This is how Kathrin Müller describes Hevelius's representations of the Moon in his *Selenographia* (1647), while also showing that the technical skill involved in crafting these representations means that their authority rested for a large part with the reputation of the scholar.[24] These observations by astronomers reflect a departure from the idea that astronomy should *only* concern itself with 'mathematical hypotheses capable of [...] calculating the observed positions of heavenly bodies [...] without regard to physical reality'.[25] The visualisations of heavenly bodies helped to establish the idea

19 On Giordano Bruno's conception of space, see Granada, 'Giordano Bruno's Concept of Space', pp. 157–78.

20 Crowe, *The Extraterrestrial Life Debate 1750–1900*, p. 9.

21 Huygens, *Cosmotheoros*. This development is much more complicated than I can show here. See for an excellent study: Crowe, *The Extraterrestrial Life Debate*.

22 Van Helden, 'Galileo and Scheiner', p. 358.

23 Müller, 'How to Craft', p. 376.

24 Müller, 'How to Craft', p. 376: 'It was an openness that was supposed to make the reader trust in the claim that this imagery was made with the best of Hevelius's knowledge and skills as well as with the best equipment available'.

25 Olson, *Science and Religion*, p. 11.

that the cosmos could not only be known by looking at it, but that the visual appearance of things in space was of interest in itself.[26]

(Re)circulating Cosmic Knowledge

Hevelius's observations of the Moon were not the only illustrations of heavenly bodies that retained their relevance for a long time. In 1666, the Royal Society of London published, in its *Philosophical Transactions*, several observations made by Jean-Dominique Cassini of the face of the planet Mars (Illustration 15).[27] The prints, which appear slightly abstracted (some contain odd dumbbell-like shapes), show spots and shapes on the surface of Mars. They are accompanied by several other observations (communicated visually and textually) made by other astronomers, most notably Robert Hooke, of the faces of Jupiter and Saturn. Cassini's representations first appeared in his *Martis Circa Axem* [...], and were also published in the *Journal des Sçavans* in 1666, highlighting the importance that was attached to the visualisation of these observations.[28]

This episode is one of the first chapters in the history of the observation of Mars – which would continue in the eighteenth and nineteenth centuries with observations by scholars such as Maraldi and Herschel.[29] However, the visualisations of these observations by Cassini also form the start of a 'standardisation' of the face of the Red Planet. It is not remarkable that the illustrations, which were based on observations of a respected scholar, were taken over by subsequent astronomers. In non-scholarly publications, too, the planets of Cassini are used well into the eighteenth century to visualise the planets of the solar system. In Nicolas de Fer's *L'Atlas curieux* (1717 ed.), for example, Cassini's renditions of the planets are used (Illustration 16). It must be noted, however, that subsequent copies generally reinforce the 'abstracted' shapes and add astrological signs to serve as a legend of sorts. Nevertheless, they highlight the ambiguity of the other shapes: do they symbolise visual features, or are they of a more symbolic nature? This particular print in the book, *Planispheres celeste*, was produced with the aid of professor La Hire, signalling a connection between scholars and publishers. In addition

26 See e.g. Müller, 'How to Craft'.

27 S.A., 'Observations', pp. 242–45.

28 'Martis Circa Axem Proprium', pp. 259–62.

29 For an excellent overview of the study (and visualisation) of Mars in this period see Sheehan, 'First Period', pp. 3–87.

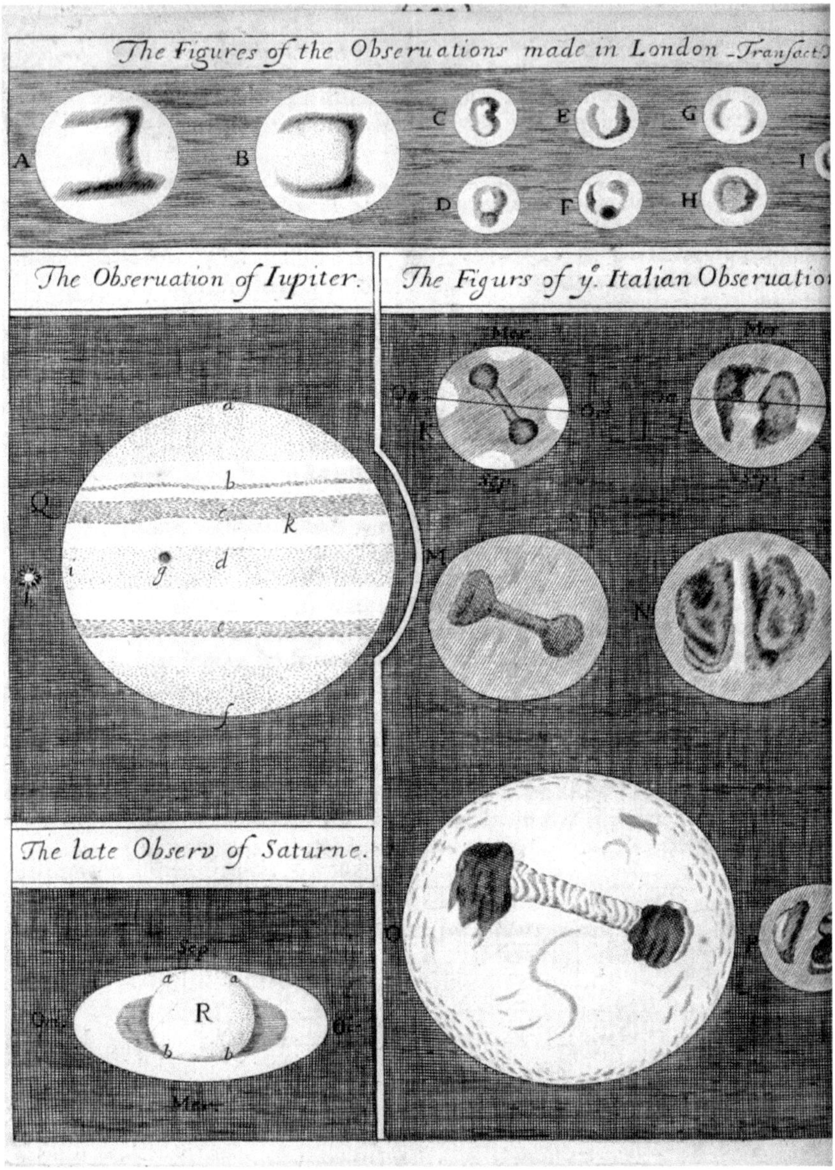

Illustration 15. Cassini's observations. The plate also depicts the observations by others. Cassini's observations are in the bottom right. Published in *The Philosophical Transactions*, vol. 1:14 (1666) (*Collection of the The Royal Society*)

to Cassini's observations of Mars, it also shows his observations of Venus, the Moon and several observations made by other scholars. De Fer's use of the illustration is far from the last, for Cassini's prints would retain their popularity into the eighteenth century. For example, in a 1790 series of prints

Illustration 16. Close up of a celestial map by Nicolas de Fer, depicting Mars according to Cassini. At the top of the map, the Sun and the Moon are depicted with references to Kircher and Cassini respectively. From Nicolas de Fer, *L'Atlas Curieux ou le Monde Represente dans des Cartes Generales et Particulieres du Ciel et de la Terre* (Paris 1717) (*David Rumsey Map Collection, David Rumsey Map Center, Stanford Libraries, list no. 12201.000*)

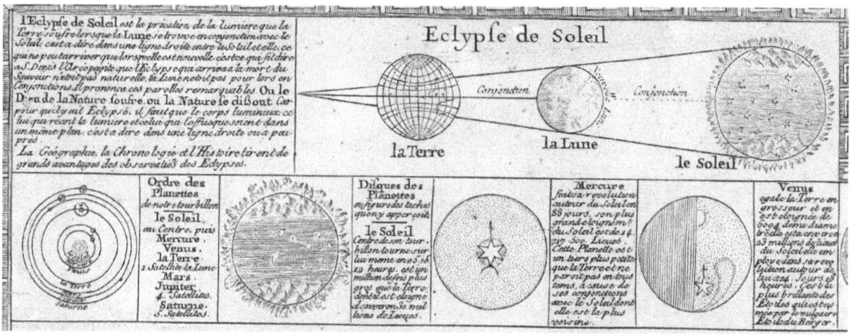

Illustration 17. Close up of a plate. From Jean-Baptiste Delafosse, *Introduction a la Geographie* (Paris ca. 1790). Hand-coloured copperplate engraving (*David Rumsey Map Collection, David Rumsey Map Center, Stanford Libraries, list no. 11954.000*)

by J.B. Delafosse, *Introduction a la geographie*, Cassini's planets feature between renditions of the orbits of those planets (Illustration 17).

It is striking how often astronomical illustrations in popularised publications are copied, and become standard ways of representing specific astronomical phenomena or heavenly bodies. A similar fate befalls the illustrations of different world systems of astronomers such as Copernicus and Brahe: in scholarly as well as non-scholarly publications, depictions

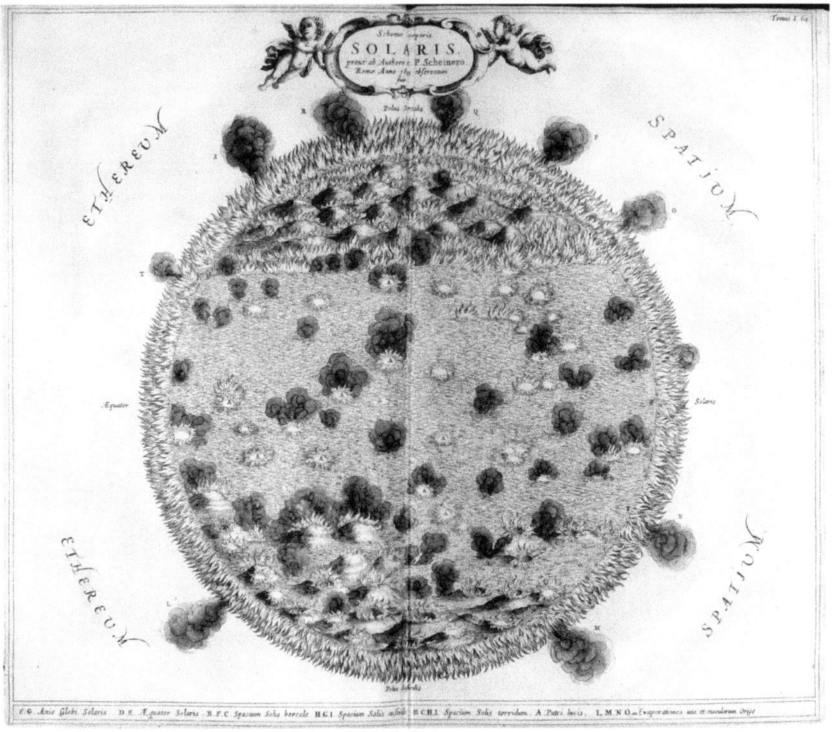

Illustration 18. Representation of the sun. Athanasius Kircher, *Mundus Subterraneus* (Amsterdam 1664–1665) (*Courtesy of The Linda Hall Library of Science, Engineering & Technology, Q155 .K6 1665*)

of their systems become, over time, symbolic of their respective theories. Particularly from the start of the early eighteenth century, we find a growing body of 'educational' prints, sold either separately or as part of textbooks, celestial atlases such as De Fer's *Atlas curieux*, but also Doppelmayr's *Atlas coelestis* or Delafosse's *Introduction a la geographie* (ca. 1790) – which are not limited to astronomy, but reflect a broader approach to knowledge about the cosmos. Many of these works are lavishly illustrated with a variety of different types of images, from diagrams and schematics to observations on the appearance of planets. Especially in the case of observations of the appearance of heavenly bodies, the copied print naturally retains much of its original. I wish to highlight two things that are of interest here. The first is the movement of visualisations from one print to the other – a type of 'canonisation' as a result of which only a selection of a few visualisations are copied over and over again. The second is the focus on the observations of the appearance of heavenly bodies. These do not only circulate in scholarly publications, but the wider public is introduced, through different kinds of prints, to visualisations of the Sun and the planets.

Illustration 19. The Sun in Allain Mallet's *Description de l'Univers*, clearly modelled after Kircher. Allain Manesson Mallet, *Description de l'Univers*, vol. 1 (Paris 1683) (*Bibliothèque nationale de France, Réserve des livres rares, G-10669*)

Cassini's views of Mars and Venus were not the only visualisations that 'hopped' from one print to the other. Other heavenly bodies, such as the Sun and the Moon befell a similar fate (Illustrations 18–20). In 1664–1665, Athanasius Kircher (1602–1690) includes a large rendition of the Sun in his *Mundus subterraneus*. Kircher's Sun is a landscape of mountains and fiery

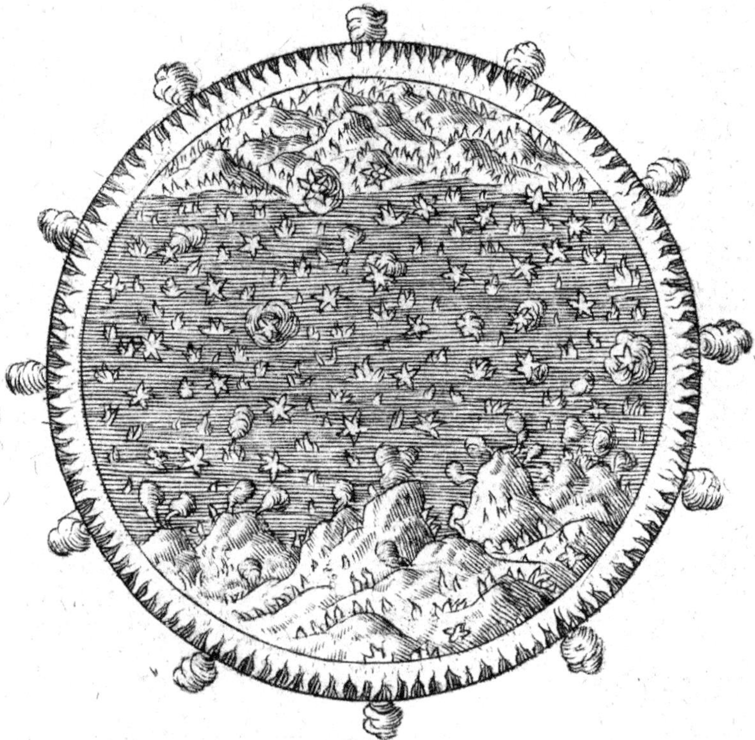

Illustration 20. The Sun in De Fer's *La Sphere Artificielle*. From Nicolas de Fer, *Atlas ou Recueil de Cartes Geographiques Dressees Sur les Nouvelles Observations [...]* (Paris 1748) (*David Rumsey Map Collection, David Rumsey Map Center, Stanford Libraries, list no. 10022.000*)

exhalations, with its own equator and climatic zones. The print is quickly taken over by virtually anyone that depicts the Sun as a heavenly body: from Allain Manesson Mallet's *Description de l'Univers* (1683) to Eberhard Kindermann's *Vollständige Astronomie*, but also in De Fer's *Atlas curieux* and Delafosse's *Introduction a la geographie*; all include a visualisation of Kircher's Sun – either identical to Kircher's original, or bearing close resemblance, highlighting Kircher's influence on this visual tradition. Sometimes, a part of the original message seems to get lost in visual translation: De Fer's 1748 rendition appears to show the Sun as a window into a fiery landscape, as if standing on the Sun, rather than as seen from Earth.

The drive to visualise the universe not only resulted in a plethora of illustrations, it also impacted the *nature* of these illustrations. This is evident

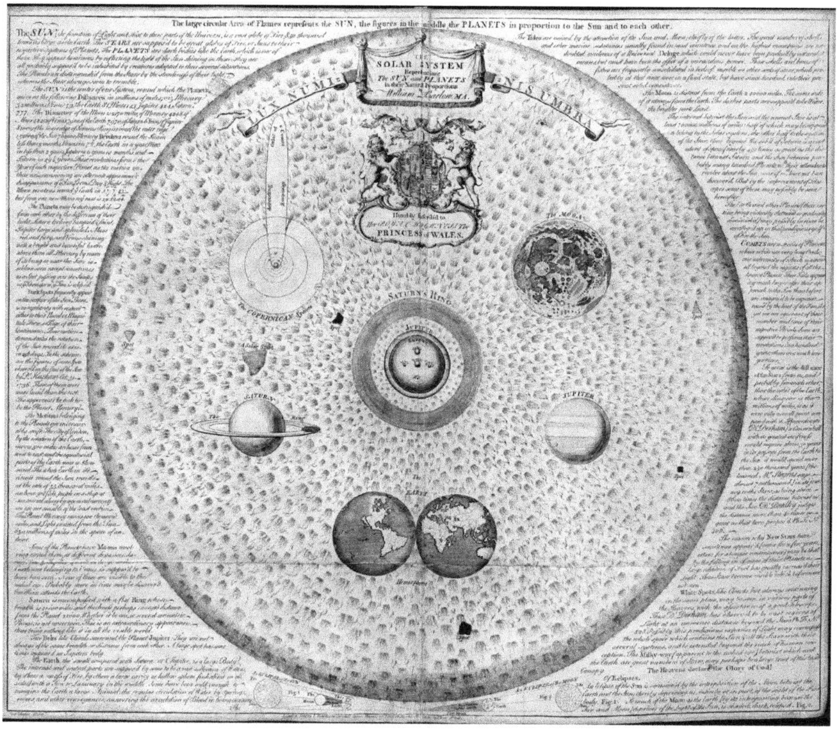

Illustration 21. Copperplate print. From William Barlow, *The Solar System Representing the Sun and Planets* (1829) (*British Library Collection, K.Top.1.52*)

from the observations of Cassini, or the many illustrations of the Moon, which focus on the visual appearance of heavenly bodies. In this impressive (and admittedly late) visualisation of the planets from 1829, the different bodies of the solar system are shown in relation to each other, so that the viewer can compare their size and appearance (Illustration 21). The text confirms what the image suggests: 'the Sun [...] is a vast globe of fire, 850 thousand times larger than the Earth [and] so great is the distance of the stars from us, and probably from each other, that the orbit of the Earth [...] were only a small point compared with it'. The reader is invited to know by seeing, while at the same time the visualisations of the planets reflect more than just their size: we can see what they would look like to the eye.

Colliding Views

Not all celestial atlases contained representations of the faces of planets – many contained mostly maps of the stars, such as John Flamsteed's

Illustration 22. Close up of Homann's *Systema Solare et Planetarium*, in which Ptolemy's system lies in shambles on the ground, and Brahe's and Copernicus's systems are described as 'according to the eye' and 'according to ratio' respectively. From Johann Baptist Homann and Johann Gabriel Doppelmayr, *Atlas Novus Coelestis [...]* (Nuremberg 1742) (*David Rumsey Map Collection, David Rumsey Map Center, Stanford Libraries, list no. 12129.006*)

posthumously published *Atlas coelestis* (1729), and stand in another visual tradition of 'star-charts'. Nevertheless, the ocular engagement with the heavens and heavenly bodies, supported by the development of the telescope and the increasing number of observation reports made by astronomers, inspired a different kind of looking, too. Observations such as those done by Cassini moved from the scholarly sphere to publications which were aimed at a wider (but still interested and necessarily relatively wealthy) public. In many cases, the representation of the appearance of heavenly bodies is closely associated with the original observer – such as Cassini, Hevelius, Hooke, and others – stressing that these observations were mediated through both skill and instruments. Apart from the diagrams and the star-charts, it would seem that a part of the attraction of these prints is that they visualise the heavens. They frame the heavenly bodies as visible entities, and make their visuality part of both scholarly and amateur understanding.

To explore the different parallel aspects of knowledge that are transmitted through astronomical illustrations, I would like to zoom in on another type of illustrations. It was not just observations of the faces of planets that migrated between scholarly and non-scholarly publications. Nor was it only in the content of the book that visual astro-knowledge is located. The 'battle between systems', as the competition between the systems of Copernicus, Ptolemy, and others is often referred to, was even fought out in visual images. Most notably in frontispieces, but also in the margins of celestial maps and educational plates. Representations of these systems contain a rhetoric that clearly favours one over the other. Volker Remmert has demonstrated that the debate surrounding the Copernican worldview gave rise to highly polemic illustrations on title pages: 'The intense reception of title iconography illustrates the stress authors laid on visual presentation and the way it could legitimize their positions'.[30] The figure of the astronomer often featured prominently in these prints, helping to establish a connection between the systems and their respective conceivers.[31] Giancarlo Nonnoi analyses the frontispiece of John Wilkins's *A Discourse Concerning a New World* (1640) with a similar result. The book, which deals with the possibility of the existence of extra-terrestrial life, is part of a popularisation of that idea. Moreover, according to Nonnoi, the book (including its programmatic and didactic title page) served to disseminate a Copernican worldview.[32]

Frontispieces were used to situate astronomical works in the discussion on world systems, but illustrations offered the author not only the chance to represent but also reflect on the several world systems by means of a visual rhetoric. In many of the celestial maps and atlases that were published over the course of this period, the author included multiple world systems in their prints – well into the eighteenth century. The continued depiction of Ptolemy's system in particular suggests a didactic or programmatic reason for that choice, as if the author wanted to place newer theories in context of older theories. In some cases, the opinion of the different worldviews is clearly reflected in their form: in the celestial map from Doppelmayr's *Atlas* (Illustration 22), for example, we can clearly see that the three world systems in the bottom right (those of Ptolemy, Brahe, and Copernicus) are presented in the order in which the author favoured them. Ptolemy's system lies in shambles on the floor, but

30 Remmert, 'In the Sign', p. 30. See also Remmert, '"Docet parva picture"', pp. 239–70. For a somewhat earlier analysis of images accompanying astronomical observation, see Elmqvist Söderlund, *Taking Possession of Astronomy*.

31 See Remmert, 'In the Sign', p. 29.

32 Nonnoi, 'Images', pp. 227–50.

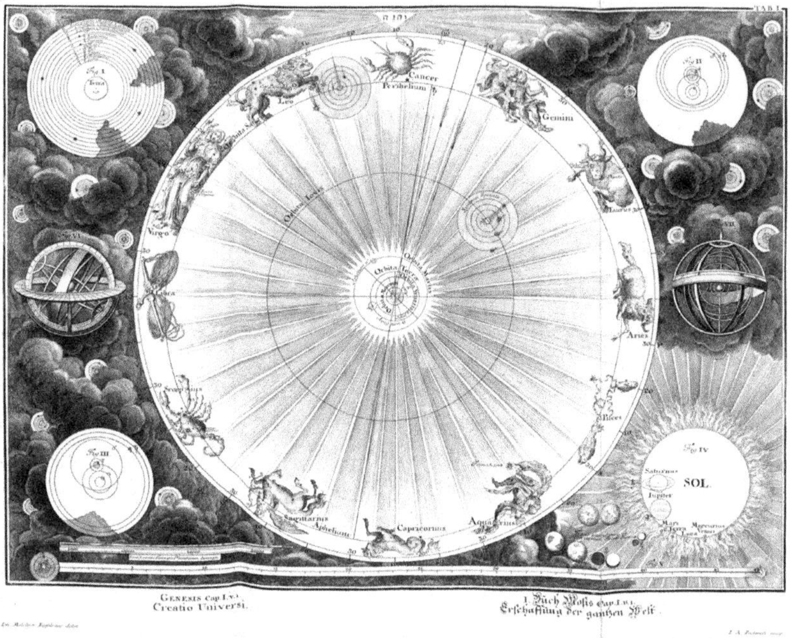

Illustration 23. View of the universe from Scheuchzer's *Physica Sacra*. Note how the clouds surrounding 'our' solar system harbour other systems, and how the three smaller world systems (such as Ptolemy's in the top left) are partly shaded. From Johann Jakob Scheuchzer, *Physica Sacra [...]* (Augsburg and Ulm 1731–1735) (*courtesy of The Linda Hall Library of Science, Engineering & Technology, QH41.S339 1731*)

Brahe's system ('sic oculis') and Copernicus's system ('sic ratione') are clearly still in some sort of competition with each other – even if Copernicus's system is presented as having the upper hand. Similarly, in Johann Scheuchzer's *Physica sacra* (Illustration 23), which copies parts of this print, the systems of Ptolemy (top left), Brahe (top right), and a merging of the systems of Brahe and Copernicus (bottom left) have shadows lurking over them, while only the system of Copernicus (in the centre) is depicted clearly. This shows the viewer what they will later read: Copernicus's is the only true system.

However, these prints do not only serve as reflections on which system should be favoured over the others. Within the representation of different systems lie different ontologies of the universe. For example, the representations of the different 'systems' of Ptolemy and Copernicus each reflect a different *type* of presentation: in the many depictions of Ptolemy's system, the areas *between* lines are those that matter: the spheres of the different planets. But in depictions of the Copernican system, as also noted by Crowther and Barker, it is the *lines* that matter, as they represent the movement of heavenly

Illustration 24. Celestial map ("Planisphærium Cœleste"). From Matthäus Seutter, *Atlas Novus [...]* (Augsburg 1730) (*David Rumsey Map Collection, David Rumsey Map Center, Stanford Libraries, list no. 12041.015*)

bodies.[33] While the different representations show stylistic similarities (both are a series of concentric circles), it is evident that they show not only different ideas of how the planets move, but also indicate that the visual elements support the representations of different ontologies of the universe. What these prints convey lies not only in what they represent, but also *how* the print represents. Conversely, knowing what they are supposed to represent changes the viewer's reading of the print and makes viewing the print dependent on knowing what it is supposed to show. Depicting, looking, knowing, and understanding are intertwined in representations of these systems.

Differences in visual languages can reflect different types of knowing. Of course, depictions of observations – diagrams, schemes, and representations of armillary spheres – are just as common. Even allegorical representations of the heavens persist, and retain at least some explanatory function. In the *Atlas novus* (ca. 1730) by Matthäus Seutter, for example, different 'visual languages' go hand in hand. The centrepiece of the print (Illustration 24) is a dual representation of the heavens, complete with a legend that distinguishes

33 See Crowther and Barker, 'Training the Intelligent Eye', pp. 453–63.

Illustration 25. 'In the beginning God created the heaven and the earth'. From Pieter de Hondt's *Taferelen der Voornaamste Geschiedenissen uit het Oude en Nieuwe Testament* (The Hague 1728) (*Collection Rijksmuseum Amsterdam, RM RP-P-1934-137*)

between different heavenly bodies, and constellations both as individual stars and as signs. The Milky Way is depicted ingeniously by means of subtle shading and coloration. However, surrounding the heavens are several smaller depictions that tell different stories. In the bottom of the print are three different 'takes' on the solar system (Brahe, Copernicus, and Ptolemy). In the top left is a cross-section of the Earth at the beginning of time,[34] in the top right a view of the Moon and the Sun.[35] Lastly, in the middle we see a series of figures: a man surrounded by seven puti. Each of these puti represents one of the heavenly bodies in the solar system, identifiable by the different attributes that they wield. The man in the centre is, of course, God – recognisable by the illuminated triangle that surrounds his head – who forms the focus of this (allegorical) universe.

These languages could also clash. In a 1728 picture Bible published in The Hague, the first passage of Genesis is represented through two views of God's Creation. Two versions of the book were published in that year: one in Dutch, one in French – both of which contain the same print (Illustration 25).[36] Looking at the illustration, the viewer might get the idea that a 'modern' understanding of the Universe, complete with Copernican heliocentrism and proto-geological ideas about the Earth, has triumphed. The orbits of the planets are not drawn as lines, but as condensed dots, suggesting a universe filled with matter. However, the text in the Dutch version suggests an entirely different reading of the print: instead of looking at the two sections, the viewer should turn to the two figures on the side. They tell the important story of how eternity and infinity came to an end, while God's Creation can only be understood through faith.[37]

I would argue that how contemporaries 'knew' the universe is constituted in part by how they saw or imagined it – knowing the universe is not only a matter of knowing how it works, but depends on the way in which this knowledge is presented, framed, communicated, and interpreted. Look, for example, at the frontispiece to Doppelmayr's *Atlas coelestis* (Illustration 26). I believe that understanding this print entails knowing what it depicts, which is dependent on how it depicts. The author and his viewers would certainly have understood that the four figures in the

34 The print is taken from Goeree, *Joodse Oudheden*, vol. 2, p. 873.

35 The text at the top reads 'God made two large lights' ('Deus faciebat duo magna luminaria') (Genesis 1:16).

36 The publication history of the work is complicated. A few years earlier, the French text by Jaques Saurin was published in Amsterdam without the print discussed here. In 1728, it was published with this illustration as Saurin, *Discours historiques*.

37 S.A., *Taferelen*, p. 1.

Illustration 26. Title page from Doppelmayer's *Atlas*. Left to right, we see Ptolemy, Copernicus, Kepler, and Brahe. From Johann Baptist Homann and Johann Gabriel Doppelmayr, *Atlas Novus Coelestis [...]* (Nuremberg 1742) (*David Rumsey Map Collection, David Rumsey Map Center, Stanford Libraries, list no. 12129.002*)

bottom half represent Ptolemy, Copernicus, Kepler, and Brahe – it says so on the labels. And they would no doubt have understood that Copernicus is correcting Ptolemy, pointing out what the universe actually was like. But to understand what goes on above the heads of these four required (and reiterated) not only knowledge of who's who and who believed what, but also knowledge of the idea that there *are* more universes, a familiarity with the mode of communicating this idea, and ultimately to have a particular kind of understanding of the universe in which one can fit this knowledge.

The Conquest of Space

My argument in this chapter is that visual illustrations do not only serve to disseminate 'dry' knowledge, but just as much work to disseminate – or at least represent – subtler aspects of this knowledge that grant it a different or additional meaning by directing the viewer's imagination. I have drawn largely from scholarly publications that date back to the middle of the seventeenth century, (celestial) atlases and other astronomical publications. In this case, the conceptual framework within which knowledge of the heavens is set is increasingly that of the universe as an empty space, governed by gravity, filled with objects we cannot just trace, but *see* through observational instruments. However, this development was decidedly not a scholarly thing alone. The reconceptualisation of the cosmos and the universe had implications for the cultural understanding of space. Starting in the late seventeenth century, the cultural imagination of space took flight, resulting amongst other things in works that have been suggested to be the first works of 'science-fiction'. Similarly, I believe that this changing conception is reflected in the visual material that was produced in this period – not only in content, but also in the perspective of, and even the very lines of these prints.

Heavenly phenomena such as comets had always attracted attention. However, over the course of the late seventeenth and early eighteenth centuries, the observation of the heavens through telescopes would become an increasingly favoured pastime of a wealthy but non-scholarly public. Astronomy was practised in some of the societies of the Republic, such as Felix Meritis in Amsterdam, where a small observatory was built on the roof of their building. At least in the Dutch Republic, astronomy in the eighteenth century was practised more extensively by amateurs (*konstgenoten*) than scholars.[38] Planetariums were a part of this development, and were both the object of public attention and a means to educate this public. For example, the English physicist John Desaguliers (1683–1744) used a planetarium

38 Cf. Zuidervaart, *Van 'Konstgenoten'*.

during a series of lectures on astronomy that he offered while visiting the Dutch Republic in 1731, for an audience of at least sixty-four. Huib Zuidervaart describes how Desaguliers used the planetarium during his lectures – aimed at a wealthy but non-scholarly public – to demonstrate both the scale of the universe as well as the fact that comets would, eventually, return to pass Earth over and over again.[39] This demonstration presupposes a conception of the universe as a vast, empty space – but also reflects the idea that this has to be demonstrated, or even experienced up to a certain degree.

The craze of the plurality-of-worlds theory was felt both in scholarly and non-scholarly literature. The idea gained impetus first with scholars Giordano Bruno and Johannes Kepler, for whom it was strongly connected to the question of whether the universe was infinite or not. The idea was not limited to scholarly circles – Cyrano de Bergerac published two books on the empires on the Moon and the Sun (published in 1657 and 1662), and Bernard de Fontenelle's *Entretiens sur la pluralité des mondes* was accessible enough to be read by a wider public. Over the course of the eighteenth and nineteenth centuries, this idea would fly high. For example, in 1752, Voltaire would imagine extra-terrestrial visits to Earth in his *Micromégas*.[40] What we see is an ever-widening universe in the literary imagination: from visits to the Moon, authors turn their imagination to all the planets of our solar system.

The expansion of the universe is reflected in visual depictions of the heavens. Representations of space from the late seventeenth century onward imagine the universe not in terms of structure, but in terms of spatiality – something that can be seen in the very lines of some of these prints. Visualisations and ontologies of space go hand in hand in the representation of different Earth systems, but many prints make the spatialisation of space visually explicit. The highly influential and often-copied image from Bernard de Fontenelle's *Pluralité des mondes* (1686) springs to mind, which represents the universe not in terms of concentric spheres, but turns it into an actual space in which planets and solar systems can be seen as if one was actually sailing past them (Illustration 27). It is not only the content of the print that matters, but the visuality of it too. Our own solar system is depicted not by drawing the orbits of planets, but instead lets them bask in the glowing rays of the Sun, which are depicted as thin lines. The universe

39 Zuidervaart, *Van 'Konstgenoten'*, pp. 73, 77–81.

40 For a good account of the plurality-of-worlds thesis in this period, see Crowe, *The Extraterrestrial Life Debate*, chapters 1 and 2.

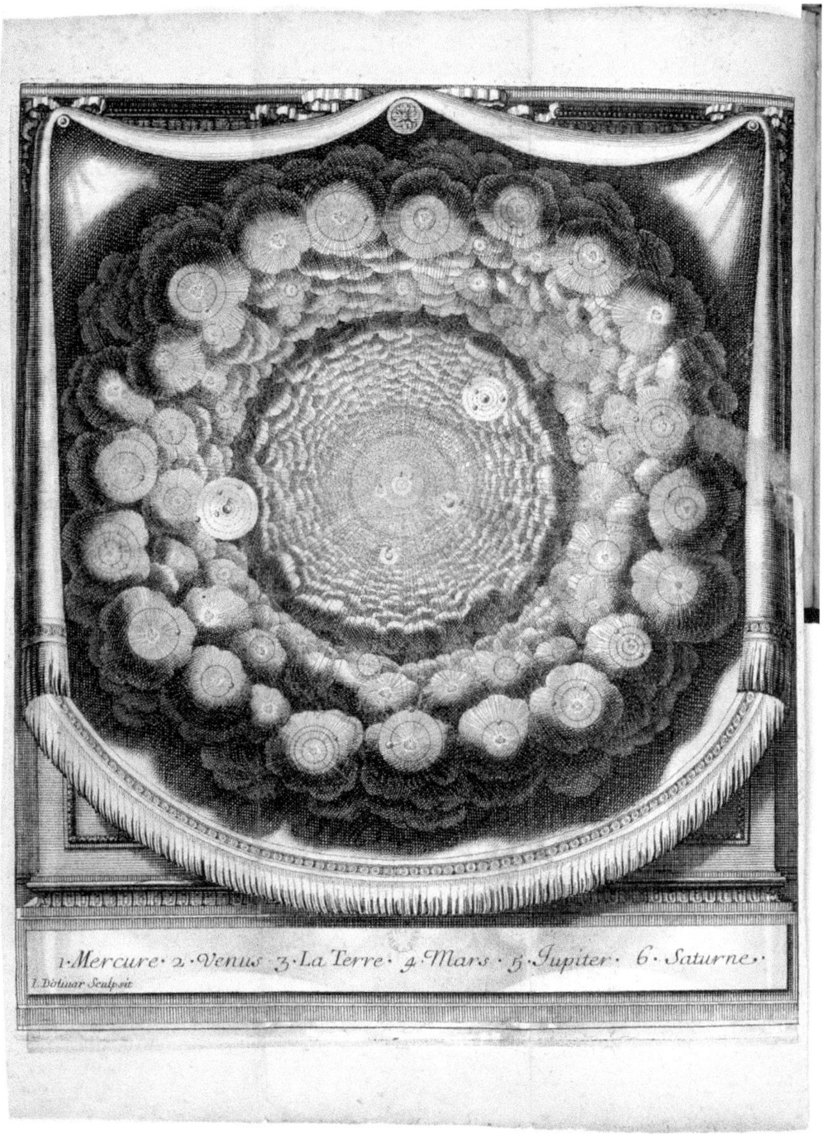

Illustration 27. Depiction of the universe. From Bernard de Fontenelle's *Entretiens sur la pluralité des mondes* (Paris 1686) (*Bibliothèque nationale de France, Réserve des livres rares, RES-R-2778*)

as a whole is portrayed as a tunnel: surrounding our own solar system are countless others, and the perspective of the print invites us to imagine seeing them all for ourselves.[41]

41 On illustrations of the plurality of worlds, see also Ayala, 'The Universe in Images'.

Another spectacular example is found in the celestial atlas by Doppelmayr from 1742 (Illustration 22). Combining different visual elements and perspectives (some of which were borrowed from others, such as Fontenelle in the top right) Homann and Doppelmayr present the spectator with a universe with shadows and depth, one that does not simply show relations or forces that operate between planets, but presents a universe in which these planets occupy real space. Fontenelle's universe serves perhaps as a reminder that the universe is infinite, but the way it is portrayed suggests more than that: solar systems are spread throughout like mushrooms in a forest, popping up everywhere from between the clouds – suggesting not a schematic perspective, but one that draws the viewer in and invites them to imagine this space. Even more telling is the centre of the print: the Sun, centre of the system, sends out its rays, casting shadows on the planets and moons that orbit around her. The scene that unfolds is not just a schematic plan, it is a view of the solar system.

This representation is copied by the aforementioned Swiss naturalist Johan Jakob Scheuchzer (1672–1733), who borrows both the rays spreading out from the Sun and the systems of Fontenelle. In his next prints, too, Scheuchzer presents us with a view of the universe that is real and three-dimensional. Scheuchzer takes the Bible, not the alphabet, as the ordering principle for his *Physica sacra* and the first section of prints details the creation of the Earth. It shows a series of depictions of the planet in which the globe is slowly brought to the fore, out of the darkness that reigned at the beginning of time. What the viewer sees is a combination of technical drawings, and visions that depict the Earth as a ball floating in space. Similarly, De Hondt's print Bible from 1728 shows both universe and Earth in two sections, one cosmic and one geological. The universe is filled not only with the orbits of planets, but with little dots that coalesce on the planets' trajectories. Reminiscent of Descartes's vortices, both Earth and heavens are teeming with matter – and both can be seen and understood in the blink of an eye.[42] (See also Illustrations 23 and 25).

In sum, this shift is a matter not only of dots and lines, but also of perspective – the viewer is presented with views of the universe that draw them in, and turn the image into a space in itself in which the viewer can imagine themselves looking at the different planets in our solar system. This perspective seems to gain popularity in the decades leading up to the publication of Scheuchzer's *Physica sacra*. With the rise of 'earth theories', which explain the creation of the Earth in mechanical and physical terms,

42 On 'cartesian cosmology' see e.g. Gaukroger, *The Emergence*, pp. 304–17.

depictions of the planet as a physical ball in space grow exponentially in number. For example, in his 1690 *Voor-Bereidsels tot de Bijbelse Wijsheid* the Dutch biblical historian Willem Goeree presents a view of the Earth without water, to illustrate its ravaged face after the Deluge. Goeree's Earth, in turn inspired by René Descartes's *Principia philosophiae* (1644), Athanasius Kircher's *Mundus subterraneus* (1664–1665), and Thomas Burnet's *The Sacred Theory of the Earth* (1684), is taken over by several authors of both world maps and celestial maps.[43] Allain Mallet's *Description du monde* (1683) presents a view of the Earth, set in a tunnel of clouds littered with stars in much the same way as the print in Fontenelle's *Pluralité des mondes* (Illustrations 28–31). The Sun illuminates the Earth with its rays, and the reader is drawn inwards into the picture.

These visual and cultural elements come together in the work of Eberhard Kindermann (b. 1715), a German astronomer and mathematician. In 1744, Kindermann published two works: a treatise on astronomy, the *Vollständige Astronomie*, and a short story titled *The Swift Journey on an Air-Ship to the Upper World* (*Die geschwinde Reise auf dem Lufft-Schiff nach der obern Welt*). Kindermann believed that he had discovered a new moon orbiting Mars, and both works explore this idea. The many, rather crudely made prints included in the *Vollständige Astronomie* paint a picture of a universe that is filled with oddly faced planets and rays extending from the Sun (Illustrations 32, 33). Kindermann's Earth is shaped like a lemon, complete with a wrinkly surface, and it floats in a universe filled with other bodies. The title page of *The Swift Journey* (*Die geschwinde Reise*) is produced in a similar style, but depicts an air-ship. The book narrates the story of five young people travelling to Mars, finding a race of long-lived aliens. Both publications exemplify that space had now been 'carved out' not only as an imaginary, but as a very real place – imagined not only in words, but also in images.

Conclusion

Understanding the heavens is not simply a matter of knowing, it is also a matter of the imagination. Over the course of the seventeenth and eighteenth centuries, the universe was spatialised: from an abstract space of incorruptible spheres, relations, and invisible forces that might not necessarily be seen as an actual 'space', scholars reconceptualised it as a concrete space with

43 On Theories of the Earth, see e.g. Ashworth Jr., *Theories of the Earth* [exhibition catalogue]. Magruder, 'Global Visions', pp. 234–57.

Illustrations 28–31. Four plates from Mallet's *Description de l'Univers* showing the earth and the solar system set in a similar tunnel of clouds as in Fontenelle's *Entretiens* (even though Mallet's representations do not depict other worlds). Nevertheless, the perspective in especially the first and last prints invites a spatial understanding of the universe. From Allain Manesson Mallet, *Description de l'Univers* (Paris 1683) (*Bibliothèque nationale de France, Réserve des livres rares, G-10669*)

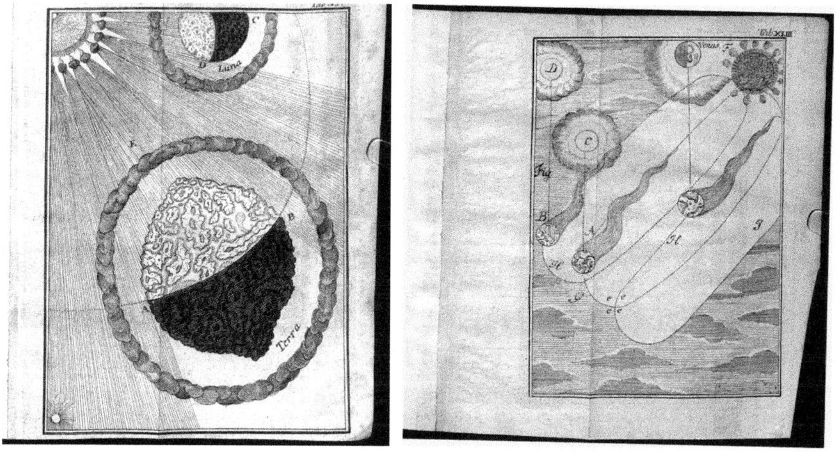

Illustrations 32, 33. Note that Venus is depicted after Cassini (see also plate 3). From Eberhard Kindermann's *Vollständige Astronomie* (Dresden 1744) (*Bayerische Staatsbibliothek München, Res/4 Astr.u. 84-1*)

visible and knowable forces. This is not just an explicit scholarly endeavour: a part of that change lies in the imagination of 'space', which is thoroughly intertwined with the visual illustrations that were made to represent it. This is not to say that it bears no relation to 'science', as this imagination of space shapes and is given shape through the work of its devotees.

The prints that I have discussed reflect an increased drive to provide its viewers with views of space – whether that is in the form of visualisations of the Moon's surface, or of a universe which they can imagine themselves in. One of the ways in which these prints communicate knowledge and information about the cosmos is by representing the universe as a vast space, filled with planets and solar systems. They confront viewers not only with representations of the orbits of planets, but bring them 'face to face' with these heavenly bodies, and engage the viewer to imagine themselves as part of the universe.

Illustrations were one of the ways through which scholarly ideas reached a non-scholarly audience. The material analysed in this chapter offers the opportunity to reflect on the different aspects or 'layers' of knowledge that are transmitted through astronomical illustrations – the information itself, the forms in which it is communicated, and the context within which it is given meaning. These prints do much more than straightforwardly convey neutral information: they function as teaching instruments, as rhetorical and programmatic devices, and they reflect (often implicitly) a subtler understanding of what they depict than is evident at first sight. Visualisations such as those in Hevelius's *Selenographia* or Cassini's publications in the *Philosophical Transactions* quickly become commonplace ways for representing the heavens, and are copied over and over both in scholarly publications as well as in those aimed at an amateur public. It shows how a more public imagination of what goes on above our head came to be. The imagination of heavenly bodies was the result of both the observations of scholars, and the dynamics that surround these illustrations moving from one 'sphere' to the other. However, the visualisations were not only used to convey knowledge about the heavens, but carry with them more subtle types of knowing, and reflect a changing perspective on the universe. One wonders if the most profound effect of these visualisations is not precisely that: they invite us to imagine our place amongst the stars.

Bibliography

Aït-Touati, F., *Fictions of the Cosmos: Science and Literature in the Seventeenth Century* (Chicago 2011).

Antognazza, M.R., 'Leibniz and the Post-Copernican Universe: Koyré Revisited', *Studies in History and Philosophy of Science* 34 (2003), pp. 309–27.

Ashworth Jr., W.B., *Theories of the Earth, 1644–1830: The History of a Genre* [exhibition catalogue] (Kansas City 1984).

Ayala, L., 'The Universe in Images: An Art-Historical Approach to the Plurality of Worlds', in: N. Campion and R. Sinclair, ed., *Culture and Cosmos* 16, no. 1–2 (2012), pp. 283–303.

Cosgrove, D., *Apollo's Eye: A Cartographic Genealogy of the Earth in the Western Imagination* (Baltimore 2001).

Crowe, M.J., *The Extraterrestrial Life Debate 1750–1900: The Idea of a Plurality of Worlds from Kant to Lowell* (1st paperback edition; Cambridge 1988).

Crowther, K.M., and P. Barker, 'Training the Intelligent Eye: Understanding Illustrations in Early Modern Astronomy Texts', *Isis* 104, no. 3 (2013), pp. 429–70.

DiSalle, R., 'Newton's Philosophical Analysis of Space and Time', in: R. Iliffe and G.E. Smith, ed., *The Cambridge Companion to Newton* (2nd edition; Cambridge 2016), pp. 34–60.

Elmqvist Söderlund, I., *Taking Possession of Astronomy: Frontispieces and Illustrated Title Pages in 17th-Century Books on Astronomy* (Stockholm 2010).

Force, J.E., *William Whiston: Honest Newtonian* (Cambridge 1985).

Gaukroger, S., *The Emergence of a Scientific Culture: Science and the Shaping of Modernity 1210–1685* (Oxford 2006).

Genuth, S.S., *Comets, Popular Culture and the Birth of Modern Cosmology* (Princeton 1997).

Goeree, W., *Joodse Oudheden ofte Voor-Berijdselen tot de Bijbelsche Wijsheid*, vols. 1–2 (Amsterdam 1690).
Granada, M.Á., 'Giordano Bruno's Concept of Space: Cosmological and Theological Aspects', in: F.A. Bakker et al., ed., *Space, Imagination and the Cosmos from Antiquity to the Early Modern Period* (corrected version; Cham 2019), pp. 157–78.
Grant, E., *Much Ado about Nothing: Theories of Space and Vacuum from the Middle Ages to the Scientific Revolution* (Cambridge 1981).
Homann, J.B., *Neuer Atlas bestehend in auserlesenen und allerneuesten Land-Charten über die gantze Welt* (Nuremberg 1707).
Huygens, C., *Cosmotheoros sive de terris coelestibus earumque ornatu coniecturae* (The Hague 1698).
Journal for the History of Astronomy 42, no. 1 (2011).
Kusukawa, S., and I. Maclean, *Transmitting Knowledge: Words, Images and Instruments in Early Modern Europe* (Oxford 2006).
Lefèvre, W. et al., ed., *The Power of Images in Early Modern Science* (Basel 2003).
Lüthy, C., and A. Smets, 'Words, Lines, Diagrams, Images: Towards a History of Scientific Imagery', *Early Science and Medicine* 14, no. 1–3 (2009), pp. 398–439.
Magruder, K.V., 'Global Visions and the Establishment of Theories of the Earth', *Centaurus* 48 (2006), pp. 234–57.
'Martis Circa Axem Proprium reuolubilis Obseruationes. Bononiae à 10. Dominico Cassino habitae. Romae, 1666', *Journal des Sçavans* 22 (1666), pp. 259–62.
Müller, K., 'How to Craft Telescopic Observation in a Book: Hevelius's *Selenographia* (1647) and Its Images', in: N. Jardine and R. Raphael, ed., *Forms and Functions of Early Modern Celestial Imagery* [special issue]. *Journal for the History of Astronomy* 41, no. 3 (2010), pp. 355–79.
Nonnoi, G., 'Images, Models and Symbols in Copernican Propaganda', in: W. Lefèvre et al., ed., *The Power of Images in Early Modern Science* (Basel 2003).
Olson, R., *Science and Religion, 1450–1900: From Copernicus to Darwin* (Baltimore 2006).
Randles, W.G.L., *The Unmaking of the Medieval Christian Cosmos, 1500–1760: From Solid Heavens to Boundless Aether* (Aldershot 1999).
Reeves, E., *Painting the Heavens: Art and Science in the Age of Galileo* (Princeton 1997).
Remmert, V.R., 'In the Sign of Galileo: Pictorial Representation in the 17th-Century Copernican Debate', *Endeavour* 27, no. 1 (2003), pp. 26–31.
Remmert, V.R., '"Docet parva picture, quod multae scripturae non dicunt". Frontispieces, Their Functions, and Their Audiences in Seventeenth-Century Mathematical Sciences', in: S. Kusukawa and I. Maclean, *Transmitting Knowledge: Words, Images and Instruments in Early Modern Europe* (Oxford 2006), pp. 239–70.
Rowland, I., 'Representing the World', in: R. Bod et al., ed., *The Making of the Humanities*, vol. 1 (Amsterdam 2010), pp. 75–105.

S.A., 'Observations Made in Italy, Confirming the Former, and withal Fixing the Period of the Revolution of Mars', *Philosophical Transactions A* 1, 14 (1666), pp. 242–45.

S.A., *Taferelen der Voornaamste Geschiedenissen van het Oude en Nieuwe Testament* (The Hague 1728).

Saurin, J., *Discours historiques, critiques, theologiques, et moraux* [...] (The Hague 1728).

Sheehan, W., 'First Period, 1636–1830', in: W. Sheehan, ed., *Camille Flammarion's The Planet Mars* (Cham 2015), pp. 3–87.

Theisohn, P. et al., 'Astronomie und Weltraumfiktion: Zur Einleitung', in: W. Fink et al., ed., *Des Sirius goldne Küsten – Astronomie und Weltraumfiktion* (Leiden 2019), iv–x.

Van Gent, R.H., 'The *Atlas Coelestis* (1742) of Johann Gabriel Doppelmayer'. https://webspace.science.uu.nl/~gent0113/doppelmayr/doppelmayr.htm [accessed 29 April 2024].

Van Helden, A., 'Galileo and Scheiner on Sunspots: A Case Study in the Visual Language of Astronomy', *Proceedings of the American Philosophical Society* 140, no. 3 (1996), pp. 358–96.

Whiston, W., *A New Theory of the Earth* (London 1696).

Zuidervaart, H.J., *Van 'Konstgenoten' en Hemelse Fenomenen: Nederlandse Sterrenkunde in de Achttiende Eeuw* (Rotterdam 1999).

About the Author

Wouter de Vries is a lecturer in history at Leiden University.

10. The Third Book of Wisdom: Eisinga, the Planetarium, and Eighteenth-Century Learning

Fokko Jan Dijksterhuis

Abstract: This chapter explores the meaning and use of Eisinga's planetarium in the context of eighteenth-century experimental philosophy, physico-theology, and the culture of improvement. It takes a cue from Eisinga himself, presenting the planetarium as a means to spend one's idle hours usefully by contemplating creation. As an analogon to the universe the planetarium thus mediates our understanding. The chapter positions Eisinga and his planetarium in the epistemic and pedagogic ambitions of (Dutch) experimental philosophy and against the backdrop of physico-theology. Within the context of late eighteenth-century Dutch society, Eisinga can be considered as a representative of a new category of knowledge practitioner, the middle-class amateur.

Keywords: technological mediation, experimental learning, knowledge history, Eise Eisinga

After Eise Eisinga (1744–1828) finished his planetarium in 1781, he wrote an elaborate description of it: 'Nauwkeurige Afteekeningen en Beschrijving' (1784).[1] The description was addressed to his sons Jelte and Jacobus, to teach them how to maintain and keep running his 'Heaven's Circus, Sun and Moon hands' after his death (Illustration 34). He remarked that he had created the

1 The description exists in two manuscript versions; their texts are identical: the 92 page EE07 and the 98 page EE13. A transcription of EE07 was published in Havinga et al., *Planetariumboek*, 253–330. The manuscript contains a note at the end that it was seen in 1843 by the curators of the Athenaeum and the governor of the province of Friesland upon transfer to the archive.

Bijleveld, Nikolaj, Arjen Dijkstra, and Samuel Gessner, eds. *Cultivating and Communicating Natural and Technical Knowledge around 1800: Devotees of Science*.
Taylor & Francis Group, 2025
DOI 10.5117/9789463728577_CH10

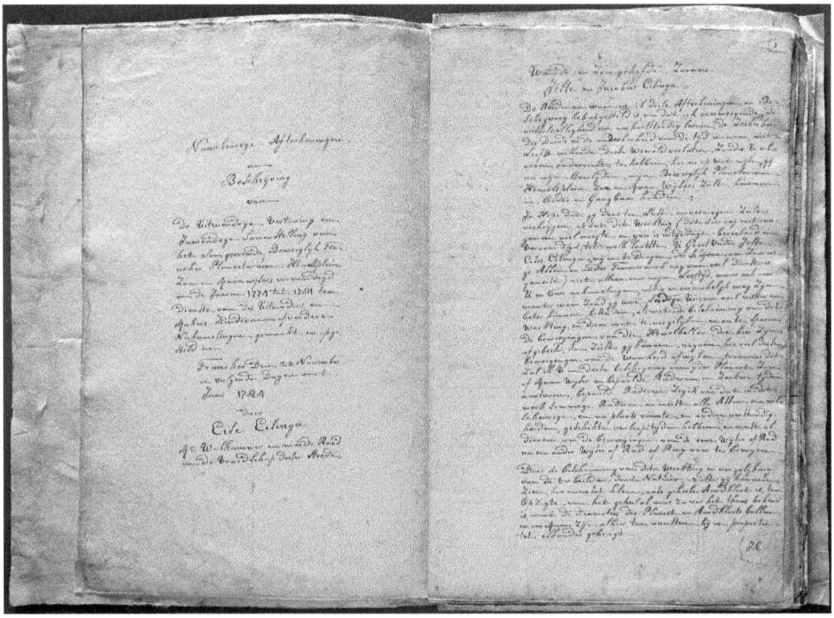

Illustration 34. Dedication to his sons, in Eise Eisinga's handwritten manual to the Planetarium. From Eise Eisinga, *Nauwkeurige afteekingen en beschrijving [...] van het [...] planeetarium*. Photo Haye Bijlstra (*Collection Royal Eise Eisinga Planetarium Franeker*)

machine together with his father, with much effort and diligence. Eisinga expressed the hope that he could pass on the appetite and ability to run the machine in order to let it be useful and entertaining for his times and for future generations. At that moment Jelte was ten years old, Jacobus had been born that Spring; Jelte would not survive his father, but we know that the planetarium would continue running.

The opening paragraphs of the description offer a rare moment of reflection by Eisinga on the meaning and use of his machine.

> [...] for what would be a better and more useful way to spend you idle hours than by consideration of this machine, and to compare and investigate the motions of those heavenly spheres with it [...]. By the consideration of this Machine and the comparison of the Nature being represented, you will be able to see how triflingly small our whole earthly globe is, in relation to the universe, as far is known now.[2]

2 '[...] want waar Zoud gij uwe Leedege Uuren veel nutter en beter kunnen besteeden, als met de beschouwing van dit werktuig, en daar met te vergelijken en na te sporen de beweegingen van die Hemelbollen, die hier zijn afgebeeld, dan Zult gij kunnen nagaan, hoe veel deese beweegingen,

With these words for his sons, Eisinga emphasised two aspects of the planetarium: it was a way to pass idle hours and a means for studying nature. After these reflections Eisinga went on to do what he was best at and dived into more than ninety pages of the technical details of his planetarium, the wheels and axes, proportions and corrections.

The fame of Eisinga's planetarium is deserved; it is a beautiful and ingenious contraption that creates a wonderfully immersive experience. Still, it also is an odd thing. It is a model of the cosmos, while the skies can be observed directly above. It was an immense investment of time and money into a non-commercial enterprise. What is the use of it all? This chapter addresses this question by discussing the epistemic and cultural uses of the planetarium. What exactly does one learn by considering a model, and how does one learn? What are 'idle hours' and what is the relationship between learning and leisure? Rather than asking about the importance and uniqueness of the planetarium, this chapter looks into its meaning for Eisinga and his times. It takes a tour through Eisinga's learning and his ideas about learning, via eighteenth-century experimental philosophy in the Low Countries, passing by some early modern ideas about pedagogy, and finishing with the role and meaning of 'idle hours' in the world of the enlightened middle-class. It shows how Eisinga and his planetarium can be understood as exponents of eighteenth-century developments in experimental philosophy, conceptions of learning, and the culture of improvement.[3]

Restoration of the University

Insight into Eisinga's ideas about science and education can be obtained from a report he contributed to in 1798, after he had returned to Franeker from his exile in Germany and Groningen. The university had been affected severely by the political upheaval of the preceding years and was in need of repair.[4] In the summer of 1797, the Makkum earthenware manufacturer

van de Waarheid afwijken. Door de beschouwing van dit Werktuig en vergelijking van de verbeeld wordende Natuur, zult gij kunnen zien, hoe onnosel klein onse geheele Aardkloot is, ten Opzichte van het geheel al, voor *zo* ver het thans bekend is, want de Diameters der Planeet en Aardkloots bollen en ons Maan zijn alhier ten naasten bij in proportie tot elkander gebragt'. Havinga et al., *Planetariumboek*, p. 253.

3 For an overview of eighteenth-century views on learning, see Van Berkel in this volume.

4 Tigchelaar and Eisinga, *Plan*. For the context, including initiatives for higher education reform at a national level, see Mijnhardt 'Onderwijs en wetenschap'. See also Dijkstra in this volume.

Yme Freeks Tigchelaar (1753–1812) was commissioned to elaborate a *Plan for Restoration of the Academy in Franeker*. Tigchelaar grasped this opportunity to include a proposal for renovating the university with a discourse on the utility of a new organisation of the academy. Tigchelaar asked Eisinga for support, in particular regarding the sciences. The latter willingly accepted the request, being 'an advocate of the sciences in general and of the citizens of his own town Franeker in particular'.[5] This shows that Eisinga had made a name for himself as knowledgeable of the sciences.

Given this background to the text, we can assume that the *Plan* reflects Eisinga's views, in particular regarding the sciences. The discussion of mathematics does indeed fit very well with his approach and activities. The *Plan* explained how mathematics makes the positions and motions of bodies comprehensible. The 'attentive youth' will realise that this also extends to the bodies beyond the earthly dominion – the Moon, planets, and stars. The story of Eisinga's life, to bring the heavens closer, finds expression in this passage:

> What appears very incomprehensible to you is this: that we Humans, staying here on our ground, are well able to do calculations and predictions that are mathematically infallibly true, about the rising and setting of the Sun and about the Moon with all its different appearances, and also about the stars that are so incredibly far from us.[6]

The *Plan* continued with a brief exposition on the teaching of mathematics – step by step – as well as on the usefulness of mathematics for reasoning in general and as a foundation of the sciences, in particular astronomy. The next section, on natural history and mechanics, was a plea for grounding the art of making machines on mathematics. In this way, the *Plan* was to bring Franeker up to date with the enlightened sciences as they had developed in the eighteenth century.[7]

5 Tigchelaar and Eisinga, *Plan*, p. 8: 'Hierom verzogt ik den Burger Eisinga tot ondersteuning, het welk mij toegestaan wierd, en die dit als Voorstander van Weetenschappen in 't algemeen, en der Burgeren zijner eigene Woonplaats Franeker in 't bijzonder, ook gereedelijk aannam'.

6 Dijkstra, *It Himelbouwerke*. Tigchelaar and Eisinga, *Plan*, p. 24: 'Het geene U het alleronbegrijpelijkst voorkomt is dit: dat wij Menfchen hier op onzen grond blijvende, en wel in onze eigene plaats over de Op- en Ondergang van de Zon en van de Maan met haare verschillende vertooningen, als ook van de Sterren, die alle zo verbaazend ver van ons afzijn, berekeningen en voorzeggingen kunnen doen, die wiskunftig onfeilbaar waar zijn'.

7 Further research is needed into the relationship of the *Plan* for the Franeker academy to educational developments elsewhere in Europe, in particular in the German countries (see Van Berkel in this volume).

The *Plan* adds another feature of enlightened science, the physico-theological conception of the study of nature as a key to the Creation and God's greatness.[8] The structure and amazing distances of the universe will be learned 'from astronomy on mathematical foundations. "What greatness of the power and wisdom of God do I discover here anew" you say to yourself, "but how radiates his great goodness through this".'[9]

The *Plan* also expressed the conviction that those distant stars are inhabited and that those creatures serve and praise God like humans do. This reasoning is reminiscent of Christiaan Huygens's *Cosmotheoros* (1698), who maintained that the planets must be inhabited in order for all parts of Creation to praise God. The idea of extra-terrestrial life was quite common in the Enlightenment. In the *Philosophische Bedenkingen* of 1774 – which allegedly prompted Eisinga to build the planetarium – Eelco Alta (1723–1798) also entertained the thought of inhabited planets, without having them admire the Lord though.[10] The conviction that extra-terrestrial life exists was based, the *Plan* made clear, on mathematics, natural philosophy, moral certainty, and the honour of the Creator and Sustainer of the universe. Such an explicit deistic perspective is absent, by the way, in Eisinga's description of the planetarium for his sons. One would expect some edifying message in a text like that, but the description only gears towards admiration for the sheer size of the solar system, not its creator. In the description the universe is just nature.

The co-production of Tigchelaar and Eisinga contained ideas on learning that were suffused with Enlightenment pedagogy. In particular the views of German educators like Johann Basedow (1724–1790) had been influential in the Netherlands and can be discerned in the *Plan*.[11] Basedow emphasised the inherent drive for learning in children and developed a playful pedagogy based on natural learning.[12] Basedow also incorporated a moral philosophy on religious foundation, a kind of Enlightenment thinking we have also seen in the *Plan*. The basis of learning, Tigchelaar wrote, was the exercise of

8 See Van Berkel as well as De Vries in this volume.

9 Tigchelaar and Eisinga, *Plan*, p. 36: 'en ook zal deeze zoo hooge Weetenschap U leeren berekenen dat deeze Sterren een verbaazend grooten afstand van ons zijn. Dit alles zult gij kunnen leeren door de Sterrekunde op wiskundige gronden. "WELK EENE GROOTHEID DER MAGT EN WIJSHEID VAN GOD ontdek ik hier in op het nieuw" zegt gij bij U zelven, "maar hoe straalt daarin zijne groote goedheid door"'.

10 Alta, *Bedenkingen*, p. 54. In an estimation of the size of Jupiterians, he refers to paragraphs 488 and 491 in Christian Wolff's 'Elementa astronomiae' in *Elementa matheseos universae*. Huygens, *Cosmotheoros*.

11 Baggerman and Dekker, *Kind van de toekomst*, pp. 68–78. Basedow, *Elementarwerk*.

12 Overhoff, *Basedow*, pp. 59–74.

reason that moulded Man's moral character and that ought to be enriched by the sciences to the benefit of society. Society on the other hand, had to facilitate education: 'The exercise of reason is completely the work of Man itself, but in its means he should receive ample support of his fellow citizens'.[13]

The *Plan* presented mathematics as the key to the exercise of reason and maintained that it had to be learned in a natural manner. How it had to be taught, the *Plan* did not really explicate. Likewise, while pointing out the importance of mathematics to the design and realisation of machines and instruments, it did not say very much about the educational role of instruments like clocks. In that regard, the instructive function of the planetarium that Eisinga had indicated in the description for his sons, was not addressed in the *Plan*.

Philosophical Teachers

Eisinga had been brought up on Enlightenment learning, in particular in the sciences. As Arjen Dijkstra has shown, his youthful notebooks drew on common textbooks in experimental philosophy. The lecture notes of his teacher Willem Wytses adopted numerous passages from Benjamin Martin (1704–1782), without simply copying them out. Dijkstra also points out the similarities between Eisinga's notebooks and the *Jongmans-onderwyzer* of Johann Knoop (1700–1769), a prominent educator at the Frisian court.[14]

From the 1730s onwards, the writer, lecturer, and instrument maker Benjamin Martin published a range of popular writings on the new natural philosophy (Illustration 35).[15] *The Philosophical Grammar; being a View of the Present State of Experimented Physiology, or Natural Philosophy* (1735) had already in 1737 been translated into Dutch as *Filozoofische Onderwyzer; of Algemeene Schets der hedendaagsche Ondervindelyke Natuurkunde*. It was published by Isaac Tirion (c.1705–1765), an Amsterdam publisher that played a key role in the dissemination of the new philosophies. Martin published a second, extended edition in 1738, that was translated for Tirion in 1744.

13 Tigchelaar and Eisinga, *Plan*, p. 7: 'De beoéffening des verstands is volstrekt het eigen-werk van den Mensch, maar in de middelen kan en moet hij veel ondersteuning van zijne Medeburgeren ontvangen'.

14 Dijkstra, *Hemelbouwer*, pp. 44–50. See also the readings of Gärtner and Palitzsch, discussed by Gluch in this volume.

15 Martin can be compared to Martin Berschitz, as discussed by Hochadel in this volume.

Illustration 35. Frontispiece from Benjamin Martin, *The Young Gentleman and Lady's Philosophy* (London 1759) (*Collection Wellcome Trust London*)

Martin used the format of a catechism, presenting natural philosophy as a question and answer between teacher and pupil. He minimalised the content of mathematics, as the translator pointed out in his preface.[16] Emphasising the 'experimented' character of the new philosophy that had liberated learning from the rooms and books of scholars, 'the main foundations of natural philosophy, could be brought into vision, and made intelligible by simple experiments'.[17] This new philosophy proved to be a reinforcement of faith, by showing the power and greatness of Creation.

In the introductory part of the *Filozoofische Onderwyzer*, consisting of four chapters, Martin laid down the conceptual and methodological basis of natural philosophy and physiology – the latter comprising somatology (matter and bodies), cosmology, aerology, and geology. This is not unlike the structure of the *Plan*. In chapter 4, Martin discussed the use of the hypotheses and the performance of experiments. Martin considered experimental philosophy the key to knowledge, presented it as a fully British product (Bacon, Boyle, Newton, Halley, Ray, Derham), and explicitly repudiated cartesian rationalism.[18] The Dutch translation added luminaries like Musschenbroek, 's Gravesande, Boerhaave, Nieuwentijt, but maintained the British orientation.[19] In this regard, the *Filozoofische Onderwyzer* was relatively outspoken in its 'Newtonian' bent.

Martin also discussed the conduct of actual, proper experiments. He said to 'prepare a large *Apparatus* of Instruments' and continued with an overview of telescopes, microscopes, barometers, thermometers, an air-pump, and so on. (Such instruments had become more common and less expensive because of the increased supply, he added.) In the second edition, Martin added footnotes with more specifications for each instrument. He also added an extra instrument: the orrery, 'which shews the Movements of all the Heavenly Bodies in the Solar System about the Sun, (by a noble Piece of Mechanism) all the Phaenomena whereof are nicely corresponding to the Truth'.[20] A footnote referred to an illustration of an orrery published by Thomas Heath that has since been lost.[21] So, at a young age, Eisinga may have heard of the idea of a planetarium, although the *Filozoofische Onderwyzer* gave little detail about the orrery. In his 1780 description of the planetarium,

16 Martin, *Filozoofische Onderwyzer*, preface [p. 5].

17 Martin, *Filozoofische Onderwyzer*, 2nd ed., preface [p. 2]. Martin himself wrote: 'All the great Points and Dogmata of the Science, here laid down and asserted, are such as have been proved by Experiments and otherwise'. Martin, *Philosophical Grammar*, p. viii.

18 Martin, *Philosophical Grammar*, pp. 18–22.

19 Martin, *Filozoofische Onderwyzer*, 2nd ed., preface [pp. 6–8]; pp. 19–21.

20 Martin, *Philosophical Grammar*, pp. 22–27.

21 Millburn, 'Benjamin Martin', pp. 378–99.

Van Swinden wrote that Eisinga had never heard or read of a planetarium.[22] He thus emphasised what he saw as Eisinga's self-made character.

It is interesting to see how the *Filozoofische Onderwyzer* presented the orrery, as a mechanism that produced phenomena that neatly corresponded to the truth. This was not unlike the description Eisinga left to his sons; the idea of a model seems to have been common in eighteenth-century experimental philosophy. In the meantime, the orrery did not return in Martin's chapter on astronomy and the solar system, nor was didactical use of the instrument explicated further. The chapter concluded by emphasising once again that knowledge of nature came from experiments rather than books, it being wonderfully curious and delightful to see all those phenomena performed by such nice instruments.[23] The orrery thus was only mentioned cursorily in the *Filozoofische Onderwyzer*, without elaborating its astronomical and didactical meaning.

Martin did discuss the details and pedagogic use of the orrery elsewhere, though. In 1760, he published a manual for his globes that included an elaborate 'Description and Use of a New Manual Orrery', specifying what could be learned from the instrument and how.[24] Martin called it a planetarium that emulated the constitution of the world and in this way provided the 'most direct and demonstrative way to knowledge'. The manual is rather technical and explains in detail how to solve problems of eclipses, phases, and so on. Martin referred to another publication where he showed the didactic use of the planetarium. The 'Young Gentleman and Lady's Philosophy' in Martin's *General Magazine* was a dialogue between a brother and sister, addressing the whole of natural philosophy. From 1763 it was published in Dutch in the *Algemeene Oefenschoole van Konsten en Weetenschappen*.[25] In the dialogue,

22 Van Swinden, *Beschryving*, p. 5: 'door eenen Man en uitgedacht, en bereekend, en vervaardigd is geworden, die alleen die kundigheeden in de grondbeginzelen der Wis- en Sterre-kunde bezit, welke men zomtyds by Liefhebbers dier Weetenschappen aantreft; die nimmer eenige handleiding tot de beschouwende en werkdadige Werktuig-kunde gehad, van niemant, hoe genaamd, eenige hulp genooten, nimmer of *Planetarium*, of beschryving, of teekening van *Planetarium* gezien, of boeken over deeze stoffe handelende, geleezen heeft?'

23 Martin, *Philosophical Grammar*, p. 28. Where Martin referred to the common ignorance of ordinary people, the Dutch translation inserted a strange derogatory remark about craftsmen: 'Welke eene onedele, laage en onkundige ziel woont 'er in de meeste Handwerksluiden en anderen van 't Gemeen!' Martin, *Filozoofische Onderwyzer*, 2nd ed., p. 29.

24 Martin, *Description and Use*, pp. 178–86 (chapter 5, with an illustration).

25 Meijer and Lublink, *Algemeen Oefenschoole*. The *Oefenschoole* was made and published by the Mennonite Pieter Meijer and the Lutheran Johannes Lublink, with contributions by the Mennonite reverend Pieter van den Bosch. Martin's *General Magazine* was a journal that appeared between 1755 and 1765. See also Dijksterhuis, 'Ways of Seeing'.

the planetarium was used to discuss questions of planetary motions and to prove the truth of the 'solar system'. In the sixth conversation, the transition was literally made by replacing the earth at the centre of the planetarium with a 'gilded sphere'. The siblings talked through the matters of astronomy, emphasising the difference between apparent motions as they are observed from the earth and the real motions of the universe. This didactic use of the planetarium was mirrored in the didactic aim of the book in a footnote that emphasised that the reader ought also to attend the accompanying plates. In this way, Martin showed how to acquire knowledge of the world by means of instruments and illustrations.

Knoop's *Jongmans Onderwyser* drew on Martin's *Filozoofische Onderwyzer*, but with an emphasis on mathematical and useful knowledge.[26] Volume one treated arithmetic, geometry, and geography, as well as book-keeping, contracts, chronology. It was practical and aimed at preparing youths for employment and business. Volume two treated philosophy (logic, physics, mechanics, architecture, astronomy, gnomonics), but again in a quite practical manner. Knoop also expounded an experimental philosophy, yet far more inclusive than Martin. Among the pioneers he also counted Descartes, Huygens, Leibniz, Tschirnhaus, and Wolff. He concluded the chapter on methodology with Martin's original list of instruments, so the orrery was not included. Continuing on the elaborate numeracy of the first volume, the second volume went into more mathematical detail of natural philosophy than Martin. The chapter on astronomy contained an additional list of instruments as well as many more calculations. With Knoop we are in Eisinga's world of the accurate determination of astronomical positions and motions.

Martin and Knoop bring us to the heart of experimental philosophy as it developed from the early eighteenth century and in which young Eisinga had been immersed. These works are representative of the main ideas of the new philosophy and the way it was popularised.[27] The philosophical teachings resonate in Eisinga's later views, as can be gleaned from the description and the *Plan*. In this regard, he fits nicely in the main trend of Enlightenment experimental philosophy. The works that can be directly linked to Eisinga did not discuss the educational and epistemic function of instruments explicitly, but other publications that were available in the Dutch Republic did. Like eighteenth-century pedagogy in general, they emphasise the didactical importance of experiments, instruments, and

26 Knoop, *Jongmans Onderwyser*.

27 Zuidervaart, 'Reflecting'.

illustrations, but do not explain how one learns. However, as I have pointed out, Eisinga's description and Martin's account of the orrery, contained interesting hints to their epistemic and educational use. The instrument is presented as a representation of nature – of the cosmos – providing special access to truth. In my view such views are telling about the development of experimental philosophy in the eighteenth century, and I would like to discuss this epistemic feature of instruments a bit further.

Mediating Instruments

Eisinga conceived his planetarium as a model with which the nature and properties of the universe could be understood. We may say that Eisinga presented his planetarium as an 'analogon', a tangible model that emulates some natural phenomenon. The concept of 'analogon' highlights the epistemic features of the instrument. Rather than replicating the essence of nature, an 'analogon' replicates the workings of nature.[28] If the effects are the same, we can say that we have found a clue towards the inner make-up and mechanisms of reality without presuming these to be represented. The planetarium can also be conceived as an 'automaton', as Martin did in his *Description*. The concept of 'automaton' does not exclude that of 'analogon', but emphasises other aspects of an instrument. On the one hand it can present ontological conceptions of nature as a clockwork. On the other hand it may highlight the role of the maker: the ingenuity of the inventor and the mastery of the creator.

The idea of instruments and machines as analogons has a long history, going back to the sixteenth-century tradition of natural magic. Here instruments are thought to channel the powers of nature. Or rather: to emulate the powers of nature, being simulacra of natural phenomena, a microcosm displaying the workings of the world. The tradition of natural magic was epitomised by the Neapolitan Giambattista Della Porta (1535–1615) and had found an influential exponent in the Low Countries in Cornelis Drebbel (1572–1633).[29] The latter's famous – or infamous if you prefer – perpetuum works in a way similar to the atmosphere acting on the oceans to create ebb and flow. In this regard the instrument offers clues to the secrets of nature.

28 Borrelli, 'Thinking'; Keller, 'Drebbel'. For an extensive account of three-dimensional models, including analogues, see Hopwood and de Chadarevian, 'Dimensions' (and other contributions to that volume).

29 Borrelli, 'Thinking'; Keller, 'Drebbel'; Dijksterhuis, 'Magi from the North'.

By recreating natural phenomena in an artificial way, the causes of these phenomena can be made known. In this way instruments as analogons become a source of natural knowledge, instead of just a tool for inquiry and making knowledge.[30]

In the experimental philosophy of the seventeenth century, instruments had become a key to acquiring knowledge about nature, as means to explore and interrogate the properties of nature. But to view instruments as a source of knowledge as such, signified a further development. The instrument was not only an instrument, a tool for testing theories, but mediated the understanding of phenomena and gave shape to ideas. Just like an experiment creates a new reality, an instrument renders a distinct phenomenon. Many of the phenomena that were created with instruments did not exist outside these experimental settings – think of vacuum, luminescence, and electricity. Several writers, in particular in German contexts, realised around 1700 that the use of instruments in experimental philosophy created new epistemic realities. Gottfried Wilhelm Leibniz (1646–1716), for one, was open to the idea that many novel phenomena did not exist outside the new, sophisticated artifacts of his time. 'Memorable things that become known through art and experiments',[31] he wrote to Ehrenfried Walther von Tschirnhaus (1651–1708). Phenomena are not so much discovered as they are created by instrumental practices. Evidently, Leibniz did not consider the constitutive role of technology in science as an impediment to the true understanding of nature, but intuited that the 'artifactual' constitution of nature and knowledge was essential to the new philosophy.

Leibniz did not systematically elaborate his intuitions regarding artefactual mediation. The question how to understand arts and artifacts as objects of understanding became a prominent theme in the works of Christian Wolff (1679–1754), in some respects a successor to Leibniz and Tschirnhaus. In his view, the rise of experiment and mathematics had transformed natural philosophy into something new, 'physik'. Still, the question remained what the role of instruments in this new *scientia* was. In the second phase of the development of his philosophical system, that began after his expulsion from Halle in 1723, Wolff tried to explicate how arts and crafts can be sources of knowledge and why they ought to be objects of study.[32] The *Discursus*

30 I discuss Drebbel's perpetuum in the context of natural magic and mechanistic philosophy in Dijksterhuis, 'Harnessing the Elements'.

31 Leibniz, *Sämtliche Schriften und Briefe*, III, 3, p. 428. 'Si qua naturae experimenta vel artis inventa memorabilia tibi innotuere, ea rogo communices'.

32 Buschmann, 'Ein Begriff für Wissenschaft und Kunst?'.

Praeliminaris of 1728 was the introduction to a series of Latin textbooks. In the chapter on the parts of philosophy, paragraph 71 discussed 'Technologie':

> A worldly wisdom of arts is possible, although this has not been pursued until now. It can be called techno-*logy* (*Kunstwissenschaft*) – technicam aut technologiam. Techno-*logy* is thus a science of arts and handiworks, or rather a science of that which has been made by men by means of the limbs of the body, the hands primarily.[33]

With this idea, Wolff went beyond the histories of trades that had been initiated by the Royal Society and other seventeenth-century institutions. He proposed a *science* of arts rather than a mere description: 'techno-logy' is the foundation of the principles of art and the artefacts realised by these.[34] In this way artefacts become a source of knowledge.

Wolff's philosophy resonated well in the Dutch Republic. His ideas tied in closely with the specific take on experimental philosophy that developed from the 1720s onwards. Like 's Gravesande, Wolff refrained from reducing science to experimental physics; in their view philosophy needed both speculative and empirical elements.[35] The physico-theological character of their philosophies was evident as well, a feature that was maintained throughout the eighteenth century. Wolff's philosophy itself was disseminated well in the Dutch Republic. From the late 1730s the Amsterdam printer Janssonius van Waesberghe (1644–1705) published an extensive translation of Wolffian works that went through several editions. For a large part of the eighteenth century Wolff's was the only complete philosophical system available in Dutch.[36] In particular in the north of the Republic, Wolff was prominent in university teaching. As early as 1728, when Wolff had just been driven to Marburg, his philosophy was taught in Groningen by Nicolaus Engelhard (1696–1765).[37] In Franeker, Samuel Koenig (1712–1757), during his tenure as professor of mathematics and philosophy, taught Wolff

33 Wolff, *Philosophia Rationalis*, p. 33: 'Possibiles quoque est philosophia artium, etsi hactenus neglecta. Eam Technicam aut Technologiam appelare posses. Est itaque Technologia scientia artium & operum artis, aut, si mavis, scientia eorum, quae organorum corporis, manuum potissium, opera ab hominibus perficiuntur'. Translation informed by German translation of Buschmann, 'Ein Begriff für Wissenschaft und Kunst?', p. 26. Compare: Wolff, *Einleitende Abhandlung*, p. 43.

34 Wolff, *Einleitende Abhandlung*, pp. 43–44.

35 Wielema, 'Leibniz and Wolff in the Netherlands'. See also Van Berkel in this volume.

36 Wielema, 'Christian Wolff in het Nederlands'.

37 Wielema, *Ketters en Verlichters*, pp. 103–9.

while emphasising the methodological similarities with Newton. His pupil Antonius Brugmans (1732–1789) followed in his footsteps, first at Franeker (1755–1766) and then at Groningen (1766–1789).[38] In this way Wolff was a benchmark of experimental philosophy in the Dutch Republic throughout the eighteenth century.

Wolff was also present in the milieu surrounding Eisinga. Besides the university professors teaching his philosophy, Wolff's name and ideas were present in many popular books. Knoop's *Jongmans Onderwyser* explicitly and favourably referred to Leibniz, Tschirnhaus, and Wolff in its presentation of experimental philosophy. Alta too, we saw, referred directly to Wolff in his consideration of planetary conjunctions, next to Newton, 's Gravesande, and others. Like Knoop's textbook, the *Plan* of Tigchelaar and Eisinga for the Franeker academy fit the Dutch, inclusive tradition in experimental philosophy.

The *Kabinet* of Nature

A lesser known, but no less interesting, representative of early Dutch experimental philosophy was Willem van Ranouw (1670–1724), a doctor and writer with a Franeker background. Ranouw was closely connected to the circles of the budding experimental philosophy in the Republic, but he fashioned his own particular accents. He expressed the idea of instruments as a source of knowledge in an intriguing way. In 1719, in Amsterdam, he started a new journal, het *Kabinet der Natuurlyke Historiën, Wetenschappen, Konsten en Handwerken* (the Cabinet of natural histories, sciences, arts, and crafts). In the preface to the first instalment, he argued that God has revealed his plan and reign through *three* books: His Word in the Holy Pages, Nature with all its wonderful creatures, and Art with all its works (produced by the incomprehensible abilities that people have received from God).[39] This idea of the arts – 'Konsten' – as a third book of wisdom was quite original

38 Wielema, *Ketters en Verlichters*, pp. 109–14.

39 Van Ranouw, *Kabinet* [n.p.]. In the dedication of the first issue: 'Want als wy begrypen, dat God onze goeden Vader door drie verscheide Boeken, zyne Heerschappy, vrywillige Regering, Grootmachtigheid, Heerlykheid, onbegrypelyke Natuur, Liefde, Goedheid, en hondert andere zaaken, dewelke wy daar door van onze lieven Schepper konnen afnoemen, heeft believen te ontdekken, en zover, als nodig is om hem aan te bidden en te gehoorzaamen, te openbaaren, [...] deze drie Boeken zyn: 1. De Heilige Bladeren. 2. De Natuur, met alle deszelfs wonderbaarlyke Schepzels. 3. De Konst met alle haare Konststukken voortgebracht door onbegrypelyke vermogens, dewelke de menschen daar toe van Godt hebben ontfangen'.

and signifies a specific development taking place in epistemic thinking in the early eighteenth century.

Ranouw explained that he had believed for a long time that natural historians had set apart nature and art too much. Both reveal nature after all: arts and artifacts stem from nature, have natural products as subject and matter. Accordingly, natural historians ought to combine the description of every natural product with an accurate description of the arts and artifacts.[40] Ranouw thus understood artifacts as embodiments of natural powers, as analogons. Studying works of art and the expertise of artisans would greatly enhance our understanding of the world and would be very fruitful for the commonweal. (He added that ordinary theologians lack proper understanding of the sciences and arts to be able to properly interpret the Bible.)

The preface to the *Kabinet* was a letter to Suffridus van Westerhuis (1668–1731), a prominent dignitary of Franeker and resident of the Martena-Huis. In his letter Ranouw recalled how he had sent Westerhuis the draft of a discourse on ancient and modern natural historians. He had never finished the book and instead had started his *Kabinet*. Ranouw originated from Brielle in the south of Holland and had relocated to Franeker as a physician in the early 1690s. In 1696 he matriculated at the university. It appears that he already had some prominence in academic circles, as witnessed by some of the laudatory poems he wrote during the 1690s. In 1700 he married Maria Boncamp (1678–after 1709), a distant relative of Westerhuis. This probably also brought Ranouw closer to Westerhuis, with whom he shared his love and knowledge of naturalia. Ranouw obtained several offices, in 1707 he was burgomaster of Franeker. In 1715 he left town, earned a degree as a medical doctor in Utrecht, and settled in Amsterdam. There he became a prolific publisher, pioneering in the new genre of spectatorial journals: the *Examinator* (1718–1720), the *Kabinet* (1721–1724), *Weekelyk Discours* (1721–1722), *Esculapius* (1723).[41]

Ranouw filled the *Kabinet* almost entirely by himself, combining his extended readings with knowledge he had acquired himself. Besides the unavoidable Boyle, Ranouw regularly referred to Wolff. The pages were filled with lengthy summaries and discussions of works in experimental philosophy, natural history, chemistry, philosophy, and of course all kinds

40 Van Ranouw, *Kabinet* [dedication]: 'Ik heb al lang geoordeelt, dat de Natuur en de Konst door de Natuur-beschryvers te ver van malkanderen afgescheiden worden; [...]'.

41 Beukers, 'De tijdschriften van Willem van Ranouw'; Leemans and Johannes, *Worm en Donder*, pp. 182–93. On Ranouw in Franeker, see also: De Crane, 'Mr. J.W. de Crane'.

of arts such as mechanics and glassmaking. The opening sentence of the first issue was, after all: God is creator and keeper of Nature, Man of Art.[42]

The first issue of the *Kabinet* contained a lengthy – somewhat babbly – exposition of the parts of natural history: the heavens, the atmosphere, the waters, and the earth. In the discussion of the atmosphere he relied on Wolff for his understanding of the nature and properties of the air, and Guericke's experiments in particular. The discussion of the air-pump brought the reader to the heights of Tenerife and Peru before returning to Amsterdam and its bad air. Here Ranouw expressed his expectation that a science of the arts would be of use to improve the living conditions in the overcrowded town. This was very much in the spirit of Dutch experimental philosophy – it was no coincidence that Wolff had named one of his early books *All Kinds of Useful Experiments* (*Allerhand nützliche Versuche*).

Ranouw's exposition on cosmology was brief and general but its approach is interesting in the light of Eisinga. He emphasised the difference between direct and mediated observation. Everyday observation of the heavens fails to reveal the magnitude and structure of the universe, the Sun, planets, moons, and stars. Once again calling on Wolff, he said:

> While our sight is too weak to discover the known diversity of these inexpressible distances. Judging by the senses solely, we cannot distinguish whether all stars are at the same distance from us.[43]

We experience the heavens as if we are at the centre of a large cosmological sphere, but it cannot be proven whether this is actually the case. Only reasoned and mediated observation can teach the structure of the heavens and our place in it. Ranouw continued with a general description of the main heavenly bodies and their (apparent) motion. Apart from this, the *Kabinet* contained little astronomy; Ranouw was primarily interested in topics relating to air and earth.[44] In his notebooks, Eisinga would also emphasise the difference between our direct perception of the heavens and

42 Van Ranouw, *Kabinet*, first issue, p. 1: 'Godt is Schepper en Onderhouder van de Natuur, en de Menfch van de Konft'.

43 Van Ranouw, *Kabinet*, first issue, p. 89: 'dewyl ons gezicht te zwak is om de bekende verscheidentheid dezer onuitdrukkelyke afgelegentheden te ontdekken. Wy konnen door 't oordeel (op de zin aanbrenging rustende) niet onderscheiden, of de sterren alle in een gelyken afstand van ons af zyn of niet'.

44 In this sense, he continued on Drebbelian interests (see above) as well as those of Fahrenheit (see Van Berkel elsewhere in this volume). Ranouw was instrumental in promoting Fahrenheit in the late 1710s.

the actual structure. This included taking into account and quantifying the effects of atmospheric refraction. Naturally, Eisinga went much further than Ranouw into the details of observations and the reconstruction of the positions and motions.

I have discussed Ranouw and Wolff quite elaborately, because their views on instruments and arts are illuminating for our understanding of Eisinga and his planetarium. Besides being representative of eighteenth-century Dutch experimental philosophy, they highlight a particular new understanding of instruments. Their ideas reflect a change from the early eighteenth century, in which instruments became a source of knowledge and acquire a mediating role in the study of nature. Although Eisinga, or his teachers, did not explicate such notions, I propose that they are key to understanding the remarkable project of the planetarium. As an analogon, the planetarium is a way of knowing the universe, rather than a mere tool to observe and study it.

In Conclusion: *Uitspanning* with *Uitspansel*

The early modern period witnessed a reconfiguring of the understanding of nature, its creator, humankind, as well as the relationships between them.[45] This reconfiguring was a multifaceted rearrangement of a variegated landscape of positions and convictions, in which we can try to locate Eisinga and his planetarium. The idea of nature as the second book of wisdom had become prominent, in particular in Protestant circles. Although it may be said to gradually disappear in the eighteenth century, it remained a leading conception in physico-theology and is evident in Eisinga's writings as well. In the meantime, the clockwork metaphor of nature had gained currency, in particular in the new philosophies of nature. When Ranouw criticised the classical distinction between nature and *technè*, he assumed some kind of technological understanding of nature, referring to nature as 'the renowned masterpiece of God'.[46] In this way the Lord becomes a craftsman inventing Nature and it is only to be expected to see the 'Konsten' as a source of wisdom. Ranouw linked this with the idea of the arts as the third book of wisdom. He probably was rather singular in explicating this idea, but it tied in with a changing conception of technology and arts as a

45 See for example, Harrison, *Territories*, in particular chapters 5 and 6; Jorink, *Book of Nature*.
46 Van Ranouw, *Kabinet*, first issue, p. 85: 'het roemruchtig konststuk Gods (dat wy Natuur noemen)'.

source of knowledge that emerged in the eighteenth century. Conceiving of instruments as mediating knowledge also has consequences for the understanding of nature.

Eisinga presented the planetarium as a model of the cosmos. As pointed out above, besides as an analogon, the planetarium can also be conceived as an automaton, a self-moving apparatus. In this regard, it quite literally expressed the clockwork metaphor. Studying the model was a way to acquire knowledge of nature and foster admiration for the wonders of the universe. In the physico-theological conception of natural philosophy, the planetarium was a way to acquire knowledge of Creation. Yet, it was also an emulation of God's Creation. Van Swinden called it a 'masterpiece representing a completely moveable system of the heavens, invented and constructed by Eise Eisinga' in his 1780 description.[47] This raises the question, what the role of the artificer was. According to Ranouw, the third book of wisdom displayed the incomprehensible abilities that people have received from God. Eisinga probably would not dare to presume that emulating the clockwork of Creation, implies that the clock-maker emulates the Lord. The Deventer minister Jacob Kuiper addressed the issue in 1782 in a 'Discourse on the Special Providence of God'.[48] The main question of the essay was whether Man is capable of recognising this providence, but as closing argument Kuiper compared Creation with a masterpiece of an artisan. The fundamental difference in his view was that God created an animated clockwork – thus enabling a perpetuum – whereas human creations were purely material.[49] Interestingly, Kuiper chose a planetarium as an example, 'an artifice that orderly indicates the course of the planets'. This seems to refer directly to Eisinga's planetarium, as it was generally known since Van Swinden's 1780 description. Moreover, Kuiper had been a minister in Makkum from 1773 to 1775, and may have had direct experience of Eisinga's project. Although it is not known whether Eisinga was aware of the possible paradox implied in emulating a clockwork universe, Kuiper's argument provides a solution for the hubris implied.

Leaving the philosophical and theological issues involved in the planetarium behind, let us turn to the artificer, the instrument maker and his place in Enlightenment experimental philosophy. Conceiving of artefacts as sources of knowledge was closely linked to the industrious character

47 Van Swinden, *Beschryving*. Compare De Vries's discussion of the visual representation of Creation in this volume.

48 It was a prize essay for Teyler's Theological Society: Kuiper, 'Verhandeling'.

49 Kuiper, 'Verhandeling', pp. 391–93.

of the Enlightenment and its culture of improvement.[50] Besides aspirations of making learning useful, this consisted of improving industry and practitioners. In the *Plan*, Tigchelaar and Eisinga pointed out the utility of the sciences for industrialists, and said that the teaching of mill design should enable diligent youngsters to discover the utility of the pursuit of mathematics and mechanics.[51] Benjamin Martin's *Filozoofische Onderwyzer* likewise maintained that 'through philosophy, artificers and craftsmen are made more capable in inventing and constructing machines that make their labour more easy and beneficial'.[52] In general, Martin spoke the language of entrepreneurs, commending investments also for learned results and using metaphors of scale increase for the blossoming of experimental philosophy. As the preface of the *Filozoofische Onderwyzer* emphasised, Martin himself had had a rural background and was an entrepreneur and devotee to the sciences. In this sense the philosophical teacher resembled Eisinga, who would have understood Martin's language well.

In line with the enlightened pedagogy of Basedow and the like, philosophical education not only had profane goals but also aimed at moral development.[53] The improvement of the intellect with sciences was aimed at personal and social utility and the promotion of moral fortune, the *Plan* stated. The profits earned by such improvements were again invested in further development, intellectually, practically, and morally. The planetarium itself is a perfect example of this. Bear in mind that Eisinga and his father had dedicated their spare time and spare money to building the wonderful contraption.[54] Instead of simply indulging in their profits, they employed them to make something useful and edifying. An instrument to learn to understand and admire the universe, as Eisinga told his sons, to spend their idle hours in a meaningful way. With this the circle is closed: the product of spare time becomes a way to spend one's leisure time.

Both the realisation of the planetarium and Eisinga's ideas of using it are illustrative of changing conceptions of time in the early modern period.

50 Popplow, 'Die ökonomische Aufklärung'; Mokyr, *A Culture of Growth*, in particular chapter 14. See also the salt industry in the northern Netherlands and the oeconomic initiatives at the Groningen university, as discussed by Van Driel, *Filthy and Fat*, chapters 6 and 7.

51 Tigchelaar and Eisinga, *Plan*, p. 34: 'Hoe nuttig zoude de Werktuigkunde zelfs zijn voor de Eigenaaren der Molenfabrieken zelve, zo tot eigen zeer groot voordeel, als tot opwekking en onderwijs van veele Braave, Naarstige en voor Weetenschappen vatbaare Jongelingen, [...]'.

52 Martin, *Filozoofische Onderwyzer*, 2nd ed., p. 8: 'De Konstenaars en Handwerksluiden worden door de Filozoofie bekwaamer gemaakt, tot het uitvinden en samenstellen van Werktuigen, die hun den arbeid gemakkelyker en voordeeliger maaken'.

53 Mijnhardt, 'Onderwijs en wetenschap'.

54 Their mechanical hobby as Halbertsma later wrote, see Dijkstra in this volume.

The growing awareness of time was reflected in the cultural prominence gained by the clock.[55] The eighteenth century witnessed an increasing distinction between labour and leisure, in particular in middle-class circles like Eisinga's. Hard work had to be balanced by recreation. In Dutch the word 'uitspanning' was used – rather than 'ontspanning', which is used nowadays. It literally means unharnessing, as one would do with horses. After a day's labour, people ought to step out and relax. However, the spare time thus created should not be wasted on vanity as the devil finds work for idle hands. Nor, to remain in the artisanal domain: an idle head is the devil's workshop.[56] As a Dutch word for firmament is 'uitspansel', the ultimate goal of the planetarium can be said to be 'uitspanning' with the 'uitspansel'.

This concludes my reflections on Eisinga's sparse reflections on the planetarium. I have tried to understand the point of the remarkable investment of time and money into a leisure project. Eisinga's planetarium can be seen as a product of its Protestant Dutch Enlightenment context. It was a leisure project intended to enlighten its users and visitors, unlike the academic orreries of someone like Almeida.[57] On the one hand, it illustrates the culture of self-improvement in the emerging middle-class. On the other hand, it expresses the main tenets of eighteenth-century experimental philosophy, in particular how it developed in the Low Countries. I have highlighted a particular feature of the planetarium, that I consider significant for understanding it. It was a model that functions as an 'analogon' to the universe. This, in my view, ties in with new conceptions of instruments as mediators of knowledge, as expressed by writers like Ranouw and Wolff. By emulating the cosmos, the planetarium created a way to understand it. The creative spirit of the artificer that was able to construct this model became highly valued in the industrious Enlightenment. Van Swinden's praise for Eisinga bears witness to this.

This leaves us with the puzzle of Eisinga himself. He was certainly devoted to science, but his scientific persona remains elusive. Unlike Moreau, he was a middle-class entrepreneur. He was not a shepherd of his community, like Alta and other pastors. He was not a populariser, like Martin or Berschitz, as he had no commercial interests in his 'konstige' expertise. The same goes for his instrument making, which remained a hobby, unlike for someone like Fahrenheit. He did not mobilise his astronomy for his business, like

55 Verhoeven, 'Tijdloos'; Baggerman 'Otto's horloge'.

56 In Dutch the phrase is 'Ledigheid is des duivels oorkussen'; literally 'idleness is the pillow of the devil'.

57 Cf. the chapter by Tirapicos and Gessner in this volume.

so many entrepreneurs in the industrious Enlightenment. He represents the eighteenth-century urge for self-improvement; which he channelled into a unique achievement: a planetarium built to scale in his living room.

Bibliography

Alta, E., *Philosophische Bedenkingen over de conjunctie van de planeten Jupiter, Mars, Venus, Mercurius en de Maan door een Liefhebber der Waarheid* (Leeuwarden 1774).

Baggerman, A., 'Otto's horloge: Verlichting, deugd en tijd in de achttiende eeuw', in: M. Grever and H. Jansen, ed., *De Ongrijpbare tijd: Temporaliteit en de constructie van het verleden* (Hilversum 2001), pp. 97–113.

Baggerman, A., and R. Dekker, *Kind van de toekomst: De wondere wereld van Otto van Eck (1780–1798)* (Amsterdam 2005).

Basedow, J., *Das Basedowische Elementarwerk: Ein Vorrath der besten Erkenntnisse zum Lernen, Lehren, Wiederholen und Nachdenken* (Leipzig 1774).

Beukers, H., 'De tijdschriften van Willem van Ranouw', *Nederlands Tijdschrift voor Geneeskunde* 125, no. 40 (1981), pp. 1613–17.

Borrelli, A., 'Thinking with Optical Objects: Glass Spheres, Lenses and Refraction in Giovan Battista Della Porta's Optical Writings', *Journal of Early Modern Studies* 3, no. 1 (2014), pp. 39–61.

Buschmann, C., 'Ein Begriff für Wissenschaft und Kunst?: Technologie bei Christian Wolff', in: U. Troitzsch, ed., *"Nutzliche Kunste": Kultur- und Sozialgeschichte der Technik im 18. Jahrhundert* (Münster 1999), pp. 23–34.

De Crane, J.W., 'Mr. J.W. de Crane aan de lezers van beide zijne verhandelingen: Herinneringen van Martena-Huis en Het aloud geslacht Martena', *De Vrije Fries* 1 (1839), pp. 277–95.

Dijksterhuis, F.J., 'Magi from the North: Instruments of Fire and Light in the Early Seventeenth Century', in: A. Borrelli et al., ed., *The Optics of Giambattista Della Porta (ca. 1535–1615): A Reassessment* (Cham 2017), pp. 125–43.

Dijksterhuis, F.J., 'Harnessing the Elements: Beeckman and Atmospheric Instruments', in: K. van Berkel et al., ed., *Knowledge and Culture in the Early Dutch Republic: Isaac Beeckman in Context* (Amsterdam 2022), pp. 369–92.

Dijksterhuis, F.J., 'Ways of Seeing: Conceptions of Visuality in Enlightenment Philosophy' in: P. Corfield and J. Norden, ed., *Media & Mediation in the Eighteenth Century* (Lund 2023), pp. 139–56.

Dijkstra, A., *De Hemelbouwer: Een biografie van Eise Eisinga* (Gorredijk 2021).

Dijkstra, A., *It Himelbouwerke* (mei yllustraasjes fan Mark Hektor) (Gorredijk 2021).

Harrison, P., *The Territories of Science and Religion* (Chicago 2015).

Havinga E. et al., *Planetariumboek Eise Eisinga* (Arnhem 1928).
Hopwood, N., and S. de Chadarevian, 'Dimensions of Modelling' in: S. de Chadarevian and N. Hopwood, ed., *Models: The Third Dimension of Science* (Stanford 2004), pp. 1–15.
Huygens, C., *Cosmotheoros* (s.l. 1698).
Jorink, E., *Reading the Book of Nature in the Dutch Golden Age, 1575–1715* (Leiden 2010).
Keller, V., 'Drebbel's Living Instruments, Hartmann's Microcosm, and Libavius's Thelesmos: Epistemic Machines before Descartes', *History of Science* 48, no. 1 (2010), pp. 39–74.
Knoop, J.H., *Jongmans Onderwyser*, vols. 1–2 (Leeuwarden 1756, 1759).
Kuiper, J., 'Verhandeling over Gods byzondere voorzienigheid', *Werken Teyler's Godgeleerd Genootschap* 2 (1782), pp. 353–424.
Leemans, I., and G.-J. Johannes, *Worm en Donder: Geschiedenis van de Nederlandse literatuur. 1700–1800: de Republiek* (Amsterdam 2017).
Leibniz, G.W., *Sämtliche Schriften und Briefe* (Berlin 1923–).
Martin, B., *Filozoofische Onderwyzer* (Amsterdam 1737).
Martin, B., *The Philosophical Grammar* (London 1738).
Martin, B., *Filozoofische Onderwyzer* (2nd edition; Amsterdam 1744).
Martin, B., *General Magazine of Arts and Sciences, Philosophical, Philological, Mathematical and Mechanical* (London 1755–1765).
Martin, B., *The Description and Use of Both the Globes, and the Armillary Sphere, and Orrery, Exemplified in a Large and Select Variety of Problems in Astronomy, Geography, Dialling, Navigation, Spherical Trigonometry, Chronology, &c.* (London 1760).
Meijer, P., and J. Lublink, ed., *Algemeen Oefenschoole van Konsten en Weetenschappen; Eerste Afdeling; Behelzende de Wysbegeerte voor Jonge Heeren en Jufferen, of, Achtereenvolgende Beschouwingen van de Werken der Natuur* (Amsterdam 1763).
Mijnhardt, W.W., 'Onderwijs en wetenschap in de Franse tijd', *Vrije Fries* 91 (2011), pp. 9–22.
Millburn, J.R., 'Benjamin Martin and the Development of the Orrery', *British Journal for the History of Science* 1973, no. 4 (1973), pp. 378–99.
Mokyr, J., *A Culture of Growth: The Origins of the Modern Economy* (Princeton 2017).
Overhoff, J., *Johann Bernhard Basedow (1724–1790): Aufklärer, Pädagoge, Menschenfreund. Eine Biografie* (Göttingen 2020).
Popplow, M., 'Die ökonomische Aufklärung als Innovationskultur des 18. Jahrhunderts zur optimierten Nutzung natürlicher Ressourcen', in: M. Popplow, ed., *Landschaften agrarisch-ökonomischen Wissens: Strategien innovativer Ressourcennutzung in Zeitschriften und Sozietäten des 18. Jahrhunderts* (Münster 2010), pp. 2–48.
Tigchelaar, Y.F., and E. Eisinga, *Plan tot herstel der Academie te Franeker, door het provinciaal bestuur van Friesland* (Franeker 1798).

Van Driel, J., *The Filthy and the Fat: Oeconomy, Chemistry and Resource Management in the Age of Revolutions, 1700–1850* [Dissertation] (University of Twente 2016).
Van Ranouw, W., *Kabinet der Natuurlyke Historien, wetenschappen, konsten en handwerken* (Amsterdam 1719–1723).
Van Swinden, J.H., *Beschryving door J.H. van Swinden van een Konst-stuk, verbeeldende een volleedig beweeglyk Hemels-Gestel, uitgedagt en vervaardigd door Eise Eisinga* (Franeker 1780).
Verhoeven, G., 'Tijdloos: Een inleiding "*in den haest*"', *Jaarboek De Zeventiende Eeuw* (2021), pp. 9–16.
Wielema, M., 'Leibniz and Wolff in the Netherlands: The Eighteenth-Century Dutch Translations of Their Writings', *Studia Leibnitiana* 25, no. 1 (1993), pp. 55–69.
Wielema, M., 'Christian Wolff in het Nederlands: De achttiende-eeuwse vertalingen van zijn Duitstalige oeuvre (1738–1768)', *Geschiedenis van de Wijsbegeerte in Nederland. Documentatieblad Werkgroep "Sassen"* 1, no. 1 (1999), pp. 55–72.
Wielema, M., *Ketters en Verlichters: De invloed van het Spinozisme en Wolffianisme op de Verlichting in Gereformeerd Nederland* [Doctoral thesis] (Free University Amsterdam 1999).
Wolff, C., *Philosophia Rationalis, sive Logica, methodo scientifica pertractata: Præmittitur discursus præliminaris de philosophia in genere* (Frankfurt/Leipzig 1728).
Wolff, C., *Einleitende Abhandlung über Philosophie im allgemeinem (Discursus Praeliminaris de Philosophia in Genere): Übersetzt, eingeleitet und herausgegeben von Günter Gawlick und Lothar Kreimendahl* (Stuttgart 2006).
Zuidervaart, H.J., 'Reflecting "Popular Culture": The Introduction, Diffusion, and Construction of the Reflecting Telescope in the Netherlands', *Annals of Science* 61, no. 4 (2004), pp. 407–52.

About the Author

Fokko Jan Dijksterhuis is Louise Thijssen-Schoute professor of Early Modern History of Knowledge at the Vrije Universiteit Amsterdam and associate professor of History of Science and Technology at the University of Twente.

Acknowledgments

In 2023, the Royal Eise Eisinga Planetarium in Franeker, the Netherlands, was listed as an UNESCO World Heritage site. It is recognised as the oldest continuously working ceiling orrery in the world. The Planetarium was created by local wool merchant Eise Eisinga in the late eighteenth century. This unique mechanism displays the real-time positions of planets.

Leading up to the UNESCO recognition, various meetings and publications were organised to connect researchers from around the world on common themes in Eisinga's background. Noteworthy activities included an event under the title 'Understanding Eisinga's Case: Non-Scholarly Practitioners around 1800' in January 2019 in Franeker, and the bid book for the UNESCO nomination from 2022. These activities focused on local history, highlighting the Planetarium's builder, alongside other case studies exploring scientific practices beyond traditional academic institutions. Valuable insights into Eisinga's life contributed to a deeper understanding of the world behind the Planetarium, shedding light on private individuals' pursuit of knowledge and their relationship with institutionalised science at the time. This led up to the first seeds and specific incentive to invite scholars to contribute to this volume.

For assistance during the process of bringing this book to completion, we thank Adrie Warmenhoven and Margriet van der Meer. For funding, we thank:

Royal Eise Eisinga Planetarium
Open Access Fund Groningen
Stifting Ta stipe fan Tresoar

Index